NEW WAYS WITH PAINT

NEW WAYS WITH PAINT

Over 100 techniques and
decorating ideas for walls, floors,
fabrics, furniture, and more.

LAUREL
GLEN

I dedicate this book to Paul Minchin who has touched my
world with endless amounts of fun and laughter and
who has brought a silent strength and peace into my life.

First published in 2001 by
Laurel Glen Publishing
An Imprint of the Advantage Publishers Group
5880 Oberlin Drive, San Diego, CA 92121-4794
www.advantagebooksonline.com

All notations of errors or omissions should be addressed to Laurel Glen
Publishing, editorial department, at the above address. All other
correspondence (author inquiries, permissions and rights) concerning
the content of this book should be addressed to
Cima Books, 32 Great Sutton Street, London EC1V 0NB.

ISBN 1-57145-329-6 (HC)
        1-57145-298-2 (TP)

Library of Congress Cataloging-in-Publication Data available
upon request.

Printed by Editoriale Johnson, Bergamo, Italy

Edited by Kate Haxell
Designed by Janet James
Photographed by Lucinda Symons
Styled by Lucyina Moody
Illustrated by Kuo Kang Chen

North American Edition
Publisher: Allen Orso
Managing Editor: JoAnn Padgett
Product Development Manager: Elizabeth McNulty
Project Editor: Bobby Wong
Assisting Editor: Mana Monzavi

c o n t e n t s

introduction

"Paint glorious paint, la la le la la la." I have sung this throughout the making of this book; just ask anyone who worked with me on it. I'm certifiably paint-mad. I have painted everything I own, apart from the obvious—I would never paint my cats. Paint can be a daunting purchase, mainly because there is such a variety of choices that you can go to pieces trying to make a decision about the all-important color or finish. However, before you race out to buy your paint, you need a plan of action. This is possibly the hardest part of the process, especially for me, as I like to feel that I'm making fast progress; but I have saved myself a lot of time and money by having a bit of self-control.

So where do you start? Does your home require a little attention or are you just tired of how it looks? Maybe you have come to realize that a room is not being used; you are eating in the kitchen because the dining room needs work and is currently a dumping ground until you decide what to do with it.

Well, you don't need to look any further for inspiration. If you're lost for ideas, I'm here to show you how to give your home a new look. If the idea of painting a wall doesn't fill you with glee, don't despair; it is all explained here with easy-to-follow

steps that will guide you through the techniques. From choosing the right paint, to picking the perfect techniques, to putting it all together in an interior, these pages make it easy.

When you flip through the pages of this book you will see how much fun I have had exploring the glorious medium of paint and how, with a little know-how, you can radically change the appearance of your home for a relatively small amount of money.

I have researched and forecasted color trends, drawing on the expertise of international colorists and color authorities. These bodies of experts decide upon not only the colors of paint but also the colors of pretty much everything else in our lives, from the colors of our cars to the color of our clothes and cosmetics. Most companies take heed of these forecasted color trends, as color has become such an important part of design.

I have made my own personal selections of what I believe is going to keep you ahead of current and future trends. Now, armed with all this information you can paint, learn, have fun, and enhance your home for years to come.

When it comes to putting a little color into your life, turn to paint. It offers the quickest and most inexpensive way of revitalizing any room in your home. There are many different types of paint to choose from and it is important that you buy the right kind for the decorating you are going to do, or everything can go horribly wrong.

In this chapter, I have looked at the varieties of paint available on the market, from the most basic, commonly available types to the more specialized paints. Each category of paint has its own section and there are sampleboards showing the different types of paint you can buy. I have described the qualities of each paint and how best to use it and to help you further when you choose a paint for a

the paints

particular job, you will find a checklist with each category telling you which surfaces the paint can be used on and which techniques it is suitable for.

Most kinds of paint can be found in home improvement stores or paint shops and at the back of this book (see page 126) you will find suppliers listings for the more unusual paints. Most of these can be bought by mail order and while they may seem quite expensive, remember that often you will be using only a small amount in a painted or stenciled detail to add a special finish to your room.

So, before you pick up your paintbrush, read this chapter carefully and take the time to plan what you are going to do and which paints you will need to make a success of your plans.

Midsheen Emulsion Paint

This paint has a slight silky finish to it for those of you who don't want a totally matte finish. Again, it is best applied with a roller and is not very forgiving on imperfect walls.

Matte Emulsion Paint

The classic matte finish; smoothly applied with a short-pile roller, this paint will look great on perfect walls. Beware, if your walls are bumpy however, every flaw will show. Try dragging or combing the surfaces instead.

Matte Emulsion Paint with Matte Varnished Stripes

For a subtle effect, use matte varnish to create designs on matte-painted walls. You can mask off stripes as I have (see page 76) or stencil designs (see page 44).

Blackboard Paint

This paint is great fun as you can paint it straight onto a wall to create your own blackboards. Try painting star- or flower-shaped blackboards in children's rooms, near the phone or on the kitchen wall for messages and notes.

Matte paints are produced to give a superb, flat, nonreflective finish to surfaces. As with all paints, the better the quality, the higher the price. I have experimented with many brands in writing this book and the old saying, "you get what you pay for" is unfortunately true. I have been tempted to buy large, cheap tubs of white emulsion, to my regret, as the paint is very watery and makes the job twice, if not three times, as long because it just doesn't cover the wall as well. But these cheap paints are ideal for splattering (see page 42).

I discovered that I can add acrylic color to base emulsion (base emulsion is what paint stores use when they mix colors up for you and can be found in any home improvement store) and mix my own colors of paint. I usually use deep-base emulsion, which looks like any other white emulsion, but in fact it contains the least amount of white pigment, so you don't have to use as much acrylic color to achieve the desired shade. However, if you want a pastel tone, add colors to a light-base emulsion and if you want a midcolor then medium-base should be used.

The only drawback to this is that base emulsion is expensive, as it is the same price with or without color added to it. I still think it is a cost-effective way of buying paint, as you can mix just as much as you need so you waste very little.

Another important lesson I learned was to mix the paint in the room that I was going to paint. So many variables will alter the color, from how much light comes into the room, to how other colors will reflect and affect the final color.

A matte finish is not confined to paint; it can also be found in water-based varnishes, which can be used to great affect if stenciled onto a wall that has been basecoated with a midsheen emulsion. The effect is subtle as the varnished areas turn a slightly darker color, creating a gentle tone-on-tone effect.

Using Matte Paints

Where?

Fabric ☑

Floors ☑

Furniture ☑
seal absorbent surfaces

Walls ☑

Windows ☑

How?

Airbrush ☑
use an air can

Colorwash ☑
mix with glaze

Comb ☑
mix with glaze

Crackle Paint ☑

Paper & Découpage ☑

Drag & Flog ☑

Freehand Paint ☑

Frottage ☑

Gild ☑

Grid Pattern ☑

Limewash ☑

Mask ☑

Monoprint ☑

Paint Mosaic ☑

Patinate ☑

Photograph & Projection ☑

Plaster ☑
as a surface finish

Roller ☑

Splatter ☑

Stamp ☑

Stencil ☑

Sponge & Stipple ☑

Woodgrain ☑
mix with glaze

THE PAINTS:
shiny

These paints come in a varying scale of gloss finishes, from silk for a soft-sheen surface to high gloss, which you can see your reflection in once it has dried. Satinwood and eggshell paints sit in between these two. Shiny paints are usually painted onto wooden skirtings, doors, and furniture, as they are hard-wearing and practical and can be washed.

The majority of shiny paints are oil-based. However, new developments are being made in paints and the new water-based acrylic shiny outdoor paints are able to withstand the punishment of the constantly changing weather just as well as the traditional oil-based versions.

Always prepare the surface carefully before painting by sanding and priming it. If you have never painted with oil-based paints before it can be a little daunting, as the consistency is a rather like molasses. For the first coat I always advise a thin and even layer of paint. Remember not to load the brush with too much paint and don't try to judge the color from the first coat, as the background primer will be showing through the brush strokes.

Apply the paint with steady upward and downward strokes rather than with criss-cross strokes, as the paint is thicker and sticky and will dribble. I have experimented with decanting the paint into a tray and using a small short-pile roller and I think that this is the most painless way of painting with oil-based paints. Keep ethyl alcohol at hand to clean your brushes or rollers.

High Gloss

This has a really shiny finish, is very hard-wearing, and looks rather like lacquer. It works well on small pieces of furniture, but you really must prepare the surface carefully, as small dents or nail holes will show and spoil the effect.

INSPIRATI

High Gloss and Satinwood Scraffito

I have painted the surface with high gloss, let it dry, and then painted a coat of satinwood over the top. While the satinwood was still wet, I used the end of a rubber-handled brush to scraffito (see page 91) a pattern through to the gloss paint. This technique is not suitable for vertical surfaces, as the paint will run.

Satinwood

For a less glossy but very durable surface, choose satinwood paint. This paint works particularly well on wooden furniture as it is less runny than high gloss, which makes it easier to paint awkward corners.

Gloss Varnish on Matte Emulsion

I painted the surface with emulsion paint and, when it was dry, painted circles of gloss varnish onto it. This gives a two-tone effect, similar to the matte varnish on matte paint (see page 10) but the varnished area is shiny.

Using Shiny Paints

Where?

Fabric	☒
Floors	☑
Furniture *seal absorbent surfaces*	☑
Walls	☑
Windows	☑

How?

Airbrush	☒
Colorwash	☒
Comb	☑
Crackle Paint	☒
Paper & Découpage	☒
Drag & Flog	☒
Freehand Paint	☑
Frottage	☒
Gild	☒
Grid Pattern	☑
Limewash	☑
Mask	☒
Monoprint	☒
Paint Mosaic	☑
Patinate	☒
Photograph & Projection	☑
Plaster	☒
Roller	☑
Splatter	☑
Stamp	☒
Stencil	☑
Sponge & Stipple	☒
Woodgrain	☒

Pearl Colorwash

This effect, used on a well-lit wall, is quite stunning. Pearlized paint was mixed with glaze (see page 20) and applied over a blue base coat using large, sweeping brushstrokes. The visible brushmarks give the finish a soft texture that suits this paint's delicate shimmer.

Printing with Pearl

Pearlized paint has been rolled onto leaves that were pressed onto the wall, leaving printed pearly motifs (see page 48). This technique has a dramatic effect, especially when used over a dark, neutral base color. The effect is subtle and possesses a fine quality, as do the leaves themselves.

Rolled Stripes

A powerful effect suitable for a feature wall. I added some orange acrylic paint to the pearlized paint so that it would sit happily with the base coat color. I masked out stripes of varying sizes (see page 76) and rolled the pearlized paint in alternate stripes over the base color.

Pearl Woodgrain

A rather unusual way of using pearlized paint is to create a woodgrain effect. Some acrylic color was mixed into the pearlized paint and applied over a dark base. A rubber comb was then rocked backward and forward across the paint to achieve the wood look (see page 68).

INSPIRATIONS

Imagine the luster of pearls, shimmering in the light and bringing a myriad of iridescent qualities to life. Think of glass beads, shells, luster china, metallic fabric: the colors, the subtlety, the infinite hues and much, much more. Then imagine being able to reproduce these qualities in your home.

Pearlized paint has all these qualities and can be used in many, many different ways. You will undoubtedly find, when you really start to explore its versatility, that this paint is almost breathtaking.

It is possible to add glamour, style, softness, and a touch of femininity to any corner of your home with pearlized paint. Use it as it is, as a soft wash, in bold stripes, or with stencils to reveal this paint's many facets, which will never bore you.

In a bedroom, pearlized paint can soften the edges and relax the mood; in a living area, it can evoke celebrity glamour; in a bathroom, it adds warmth and detail. Ultimately, whether you use it on a small area or over a large expanse, it will always go beyond your expectations.

You can use this medium on almost any surface—walls, fabric, doors, furniture, glass—as long as it is the type of pearlized paint designed for use on that surface.

I love pearlized paint, not only because its possibilities are endless, but because it delights the eye and is always surprising me. I am continually finding new ways to use it and would definitely describe myself as a pearlized paint fanatic.

Using Pearlized Paints

Where?

Fabric ☑
use pearlized fabric paint

Floors ☒

Furniture ☑
seal absorbent surfaces

Walls ☑

Windows ☒

How?

Airbrush ☑
use an air can

Colorwash ☑
mix with glaze

Comb ☑
mix with glaze

Crackle Paint ☒

Drag & Flog ☑

Freehand Paint ☑

Frottage ☑

Gild ☒

Grid Pattern ☑

Limewash ☒

Mask ☑

Monoprint ☑

Paint Mosaic ☑

Paper & Découpage ☒

Patinate ☒

Photograph & Projection ☑

Plaster ☑
as a surface finish

Roller ☑

Splatter ☑

Sponge & Stipple ☑

Stamp ☑

Stenciling ☑

Woodgrain ☑
mix with glaze

Silver Paint

This paint, like all the metallic paints, can be used to paint a whole wall. For a smooth effect you will have to use a roller as brush marks show up very clearly, as here.

Gold and Rich Gold Paint

These paints are also brilliant for adding details to painted walls, for example, with stenciling (see page 44) or splattering (see page 42).

Copper Paint

An interesting way to use this paint would be to paint an item, a flowerpot say, and then do the verdigris patination technique (see page 35) over the top of it for a really rich look.

Black Reactive Metallic Paint

This is very strong and would be very overpowering if used on a large area, so it is best to confine it to details. You simply apply a quite thick but even layer of the paint with a brush and leave it to dry completely. Turn the paint black by spraying the aging solution over the top of the paint and watch the effect develop as it dries. A coat of varnish can help to seal and protect the patina.

Green Reactive Copper Paint

This green patina is more colorful than the black and could look spectacular used on a feature wall. It is applied in exactly the same way as the black patina and the more of the aging solution you spray on, the richer the color will be.

Rust Reactive Iron Paint

This is the fastest and most spectacular of the patina paints and, again, it is applied in just the same way as the black patina. If you love these paints, paint a rusty table to stand in front of a verdigris wall; anyone coming into your room will be completely deceived into thinking that you have spent a lot of money on expensive metalwork.

Metallic paints have become widely available and are wonderfully versatile paints, mainly because there are no hard and fast rules as to how you should use them. You can go as crazy as you like. Try covering a whole wall in silver, so when light shines across it, it looks fabulously dazzling. The effect is rather like a highly polished metal surface.

If you prefer a refined, more controlled look, simply use small amounts of metallic paint to add shimmering finishing touches. Alternatively, mixed with some glaze and colorizer or acrylic, metallic paint can be quite magical when used as a colorwash over plain old emulsion paint.

I have also experimented with some new metallic paints that contain real metal. They are called reactive metal paints and can be used to imitate rusty iron, copper verdigris, and other finishes. The paint is physically very heavy because it contains real metal and is usually sold as a kit composed of the paint itself and an aging solution, which is sprayed on over the paint once it is dry to create the distressed effect. For the most authentic look, I find it best to apply two coats of the paint before spraying on the patina solution. The rust paint especially has a very pungent smell, so don't be alarmed!

If you like this metal look, do try these paints out, you will not be disappointed. You could easily go crazy for metal, as they are simple to use and the results unbelievable.

Using Metallic Paints

Where?

Fabric ☑
use metallic fabric paint

Floors ☒

Furniture ☑
seal absorbent surfaces

Walls ☑

Windows ☒

How?

Airbrush ☑
use an air can

Colorwash ☑
mix with glaze

Comb ☑
mix with glaze

Crackle Paint ☑

Drag & Flog ☑

Freehand Paint ☑

Frottage ☑

Gild ☒

Grid Pattern ☑

Limewash ☒

Mask ☑

Monoprint ☑

Paint Mosaic ☑

Paper & Découpage ☒

Patinate ☒

Photograph & Projection ☑

Plaster ☑
as a surface finish

Roller ☑

Splatter ☑

Sponge & Stipple ☑

Stamp ☑

Stencil ☑

Woodgrain ☑
mix with glaze

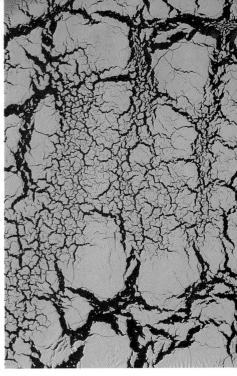

Colored Spray

There are various types of spray available, it is important to choose the right one for the job, so check the label carefully.

Stonefleck Spray

This effect comes from two coats sprayed from just one can. You really don't need any skill, just a finger to press the button on top of the can.

Crackle Effect

For this effect you need two different cans, but they are usually sold as a kit. One can provides the basecoat and the other the crackled topcoat. The thicker the basecoat, the larger the cracks in the topcoat will be.

I N S P I R A T I

Glass Etch

You can create designs on glass by masking off areas with tape or sticky-backed plastic and then spraying the glass. Here, I spray-glued a real leaf to the glass as a mask. It is an etch-effect as opposed to an acid-etch, and is not suitable for surfaces that are in contact with food or are washed often.

O N S

Why brush-paint something if you can spray it? At last manufacturers have come to realize that we would love to have beautiful homes and gardens, but we don't want to spend all our free time painting them. Hence the spray can; what could be simpler?

Two important points to mention are the fumes (and I'm not being dramatic when I say that they are smelly); you have to spray in a very well-ventilated area. The other point is to make sure you cover floors and furniture in the room, because the spray paint carries further than you think. If at all possible, spray outside, but remember to cover the grass unless you want colored grass for a while!

Spray paints are available for a multitude of surfaces from garden fences and terra cotta to stone and metal. There are transparent glass colors and glass etch for windows and glass objects or crackle-effect in a can. Colored sprays are a must, and they also come in pearls and metallics, which can be sprayed on top of colored sprays to add luster. Ceramic surfaces can be sprayed, but you shouldn't spray things that are used for food. Spray or stencil glue is always handy; I keep it for attaching stencils to awkward surfaces.

When all is said and done, the spray can is a great invention for applying a flat color quickly and you can spray on a coat of varnish from another can to seal your work.

Using Spray Paints

Where?

Fabric	☑
Floors	☒
Furniture *seal absorbent surfaces*	☑
Walls	☑
Windows	☒

How?

Airbrush	☑
Colorwash	☒
Comb	☒
Crackle Paint	☑
Paper & Découpage	☒
Drag & Flog	☒
Freehand Paint	☑
Frottage	☒
Gild	☒
Grid Pattern	☑
Limewash	☒
Mask	☑
Monoprint	☒
Paint Mosaic	☒
Patinate	☒
Photograph & Projection	☒
Plaster	☒
Roller	☒
Splatter	☒
Stamp	☒
Stencil	☑
Sponge & Stipple	☒
Woodgrain	☒

Glazes and their accompanying colorizers offer the most wonderful way of applying subtle textures and patterns onto walls. The translucent, satin finish reflects light and gives surfaces the soft sheen seen in seashells, luster-ware ceramic or silk fabric.

There are two types of glazes, those that you mix with colorizers and those that you mix with emulsion paints—do not attempt to mix the two. The first type of glaze is a milky liquid, the consistency of thick cream, to which you add colorizer which is a very densely colored pigment. You only need a tiny amount to create a very strong color, but however strong the color is, you do not lose the translucency. You can blend colorizers to make your own shades—a drop of yellow and a drop of blue will produce green. The colored glaze will remain wet and workable for longer than ordinary emulsion paint and is the perfect medium for time-consuming techniques such as flogging or stippling.

The second type of glaze is similar in appearance, but can only be mixed with emulsion paint. You mix approximately one-third glaze to two-thirds emulsion and stir the two together well. The emulsion doesn't look any different once the glaze has been added, but it will keep the paint flexible for longer.

Glazes are best used on sealed plastered or plaster-boarded walls and on prepared wood or MDF (medium density fiberboard). If you are using it on furniture, you will need to seal your work with water-based varnish. I use glazes, with colorizers or emulsions, a great deal, as they allow me to work on a surface for longer to achieve a look. I love the translucent colors and I especially love the fact that I can create exactly the shade I want to complement fabrics or other paint finishes.

Bamboo

To create this bamboo effect, I painted the wall with glaze mixed with just a little colorizer using a paintbrush. I dragged over the paint with a softening brush to blur the strokes. I cut a piece of corrugated cardboard to the width I wanted the bamboo to be and ripped off one end, across the corrugations, to create a softened edge. I dampened the torn edge (to stop it lifting off all the paint) and then dragged it in stop-start vertical lines through the paint to make the stalks of bamboo.

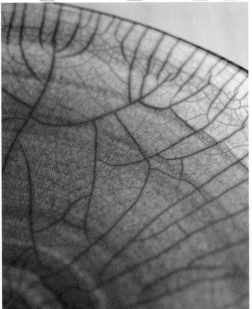

Using Glazes and Colorizers

Where?

Fabric	☒
Floors	☑
Furniture	☑
seal absorbent surfaces	
Walls	☑
Windows	☒

How?

Airbrush	☒
Colorwash	☑
Comb	☑
Crackle Paint	☒
Paper & Découpage	☒
Drag & Flog	☑
Freehand Paint	☒
Frottage	☑
Gild	☒
Grid Pattern	☑
Limewash	☒
Mask	☑
Monoprint	☑
Paint Mosaic	☑
Patinate	☒
Photograph & Projection	☒
Plaster	☑
as a surface finish	
Roller	☒
Splatter	☒
Stamp	☑
Stencil	☑
Sponge & Stipple	☑
Woodgrain	☑

Dragged Finish

Using a paintbrush, paint the wall with glaze mixed with colorizer. Smooth out the brush strokes by going over them with a softening brush. Drag a varnish brush (which is thinner than a paintbrush) through the glaze in vertical lines to create the dragged effect (see page 56).

Swept Color

This was done in exactly the same way as the bamboo effect, but I used a much bigger piece of cardboard and cut the edge of it for a sharper effect. I just swept it through the paint (see page 50).

O N S

Ragged Finish

This technique uses emulsion paint mixed with glaze. Paint the emulsion-glaze onto the wall with a paintbrush and soften any brushstrokes with a softening brush. Scrunch up a piece of rag (an old cotton towel is ideal), into a fat sausage shape and roll it through the paint, working from bottom to top. Ragging will make your hands dirty, so wear protective gloves if you wish.

Denim Paint

This a paint with a slightly gritty texture, which when applied using the dragging technique (see page 56) gives a perfect denim look. This red denim paint was dragged in two directions for a well-worn look that imitates the weft and warp of fabric.

Texture is my second love affair, though it is sometimes a closely fought battle between that and my first love—color. How can anyone live without texture? Texture plays a huge part in what I do as a designer and a large part in this book.

Texture brings colors alive and is important in making a room feel larger. I must admit that I now find it hard to paint anything a flat color, unless that very flatness serves to enliven or enhance another texture. Texture stimulates the senses and we underestimate how important it is in our lives. The idea is getting us to touch and pick up as we understand an object by touching and feeling its surface.

When creating a textured surface there are a few things to consider. Firstly, you must use the right sort of paint; an ordinary emulsion just won't work. It will dry too quickly and is too thin to hold the texture you are trying to achieve.

Secondly, if you don't use the right brushes, texturing a wall will become an ordeal rather than a fun experience. A narrow household paintbrush just won't cover enough of the surface at a time—you need a really wide brush. These are commercially available, so do get one before you experiment with any of these effects.

Finally, if you are going to cover a whole wall with one of these textured effects, do recruit a friend to help you. If you work as a team with one of you rolling the paint on and the other texturizing it, you can cover a large area quite quickly.

Cord Paint

This is a thick paint, which is even grittier than the denim paint. Roll it on over a similar colored base coat, then brush over the wet paint with a stiff, short-bristled brush to create the cord texture. I think that this paint is best used on a feature wall or in an alcove rather than all over a room.

INSPIRATI

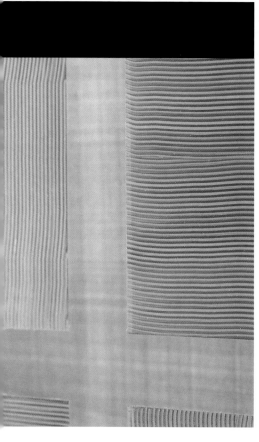

Granite Paint

For a really convincing stone finish in a room, use this granite paint. It is quite expensive, so use it on a feature wall or above a fireplace. You simply paint it on with a brush, using small brushstrokes, and the effect appears like magic.

Gel Paint

A thick, smooth paint, this can be used on glass and ceramic surfaces. It comes in various different finishes, including the translucent finish used here, which is particularly suitable for glass. To get this ridged effect, brush the paint onto the glass, then comb it with a metal comb (see page 50). The wall behind this glass has been painted in blue denim.

Artex

This is a very useful material when it comes to covering walls. The basic color is white, but you can tint it with colored powdered pigments. Don't be tempted to use liquid colorizers as these react with the artex and make it unusable. You can roll it onto a wall, but I prefer to trowel it on like plaster (see page 36), as I think it looks much better.

Using Textured Paints

Where?

Fabric	☒
Floors	☒
Furniture	☒
Walls	☑
Windows	☑

How?

Airbrush	☒
Colorwash	☒
Comb	☑
Crackle Paint	☒
Drag & Flog	☑
Freehand Paint	☒
Frottage	☒
Gild	☒
Grid Pattern	☑
Limewash	☒
Mask	☑
Monoprint	☒
Paint Mosaic	☒
Paper & Découpage	☒
Patinate	☒
Photograph & Projection	☒
Plaster	☒
Roller	☒
Splatter	☒
Sponge & Stipple	☒
Stamp	☒
Stencil	☒
Woodgrain	☒

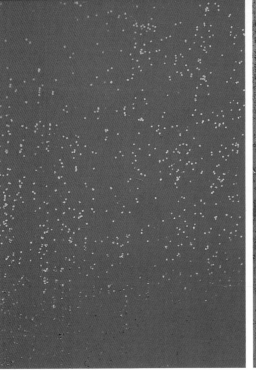

Fine Glitter

This fine glitter is suspended in a tinted, translucent base and is available in different colors. Paint it onto a wall emulsioned in a toning color or white. This is quite expensive and so is best used in small areas.

Bold Glitter

For a more intense effect try these larger glitter flakes, which are suspended in a clear base so you can paint over it in any color of emulsion. This would look great on a feature wall.

Holographic Foil

This is applied rather like transfer leaf for gilding (see page 64), but you must use a special aqua size. It is thicker than transfer leaf and can be cut to make shapes before sticking it onto a surface.

Metallic Paint

There is a section in this book devoted to metallic paints (see page 16), but they do deserve a mention here as they have reflective qualities. Metallic paints are available in various colors and can work particularly well on small areas, or when used to stencil or in stamp designs.

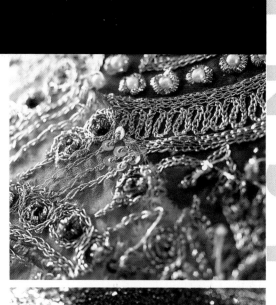

As the name suggests, reflective paints reflect the light. If you are attracted by all that sparkles, then you are in for a real treat as glitter paints, reflective foils and metallic paints can make your dreams come true.

If you imagine where the light falls upon an area of wall and treat that area with a reflective paint, the effect will be spectacular.

Glitter paint is a new paint that is applied as a topcoat to an already painted wall. The glitter particles are suspended in a mixture that allows the application of a thin and even spread of glitter with each brushstroke. It will glitter like star dust on a wall and an added appeal is that when you walk past, the glitter will appear to follow you as the sparkly surface catches the light then disappears from view. At first, this effect might sound rather eccentric, but it is one that can be as bold or as subtle as you wish. The darker the background color, the greater the contrast between the glitter and the paint. Alternatively, paint glitter onto a light ivory wall for a subtle shimmer as the light dances across the surface.

The metal foil comes in various finishes; gold, matte silver, copper, blue, or the bright silver holographic finish shown here, which in the right setting is breathtaking. The visual illusion is of a highly dazzling holographic effect encased in thin foil.

Using Reflective Paints

Where?

Fabric	☒
Floors	☒
Furniture	☑
seal absorbent surfaces	
Walls	☑
Windows	☒

How?

Airbrush	☒
Colorwash	☒
Comb	☒
Crackle Paint	☒
Paper & Découpage	☒
Drag & Flog	☒
Freehand Paint	☑
Frottage	☒
Gild	☒
Grid Pattern	☑
Limewash	☒
Mask	☑
Monoprint	☒
Paint Mosaic	☑
Patinate	☒
Photograph & Projection	☑
Plaster	☒
Roller	☑
Splatter	☑
Stamp	☒
Stencil	☑
Sponge & Stipple	☒
Woodgrain	☒

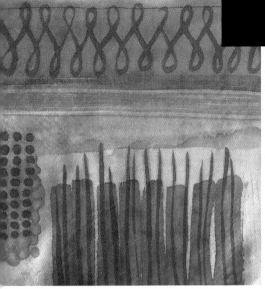

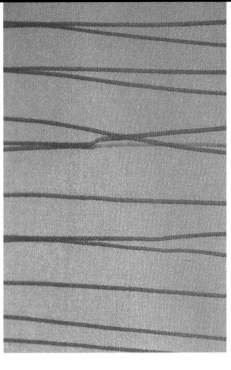

Hobby Paints

These are the most widely available fabric paints and you can buy them in small quantities. They are a bit like water-color paints for fabric; you can dilute them just a little with water and you can mix colors together to create your own shades. This is great if you are on a tight budget, as you don't have to buy all the colors. Hobby paints are best used on white, fine fabric such as silk or thin cotton. Once you have finished painting, it is vital to leave them on the fabric for as long as possible (48 hours is best) before ironing them to fix them. Always wash the painted fabric by hand.

Marker Pens

For instant results, fabric marker pens are fantastic. They tend to work best on smooth fabrics such as fine silks and T-shirt fabric and not so well on textures such as velvet or cord. Like hobby paints, the color of the background fabric will affect the color of the paint. So, if you draw on blue fabric with a purple marker pen, the result will be a brown line.

Pigment Paints

These are good fun; they are a concentrated liquid color that you add to a binder to make paint. You can use them to paint onto any color of fabric by adding white pigment to make your chosen color of paint opaque, as I did here. Adding silver or gold paint to a color will give a lustered effect. This paint sits on the surface of the fabric, rather than sinking in like hobby paints or marker pens, so it is important not to apply it too thickly. I find that the best techniques for applying this paint are monoprinting (see page 86) and stenciling (see page 44). Avoid very heavily textured fabrics like thick cord or long-pile velvet, but other than on these, the paints are very versatile.

INSPIRATI

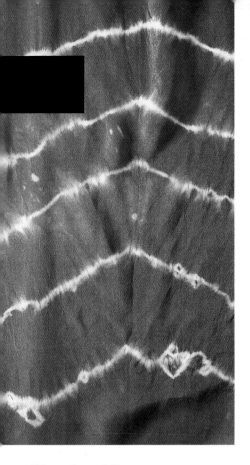

I originally trained as a textile designer, specializing in printed fabrics and my fascination with fabrics and printing has not diminished with time; if anything, my appetite for them has increased. I have quite a stack of fabric in my studio, just waiting for when the mood takes me to create something simple, unusual, and of course, all mine. It is very satisfying when someone asks where a cushion, a cover or even a skirt came from, and you can say, "I made it."

There are several different sorts of fabric paints available and it really is important to choose the right one for the project you are going to do. The colors of hobby paints, marker pens and water-based dyes will all be affected by the color of the fabric you paint them onto, so do experiment on a scrap piece before starting work on a pair of curtains or a duvet cover.

My personal favorites are the pigment and puff paints; I have to get my hair-drier out to dry them as quickly as possible because I get so impatient to see the finished result.

Water-based Dyes

If you can work quickly, you can paint designs onto fabrics with these dyes. However, once you have mixed it up, the color can only be fixed for 30 minutes, so if you are doing a big painting project, mix up small amounts at a time. Another way, and probably the best way, to use these dyes is to dip fabric into them. I made this tie-dye sample by bunching the fabric up, tying it with string and then immersing it in dye.

Puff Paint

This is a really odd paint, deceptively odd as you have no idea what the finished result will look like until you iron it and it puffs up, distorting the fabric. The basic color is white, but you can tint it with the pigment color used in pigment paint (see opposite) to make any shade you want. The puff effect works best on fine fabrics as the fabric distortion creates wonderful effects. Stenciling the paint on with a stippling brush (see page 44) is the best way to apply it. You don't have to put on a thick layer, but it should be even and you must iron it on the back within 30 minutes of it drying.

Where?

Fabric	✓
Floors	☒
Furniture	☒
Walls	☒
Windows	☒

How?

Airbrush	☒
Colorwash	☒
Comb	✓
Crackle Paint	☒
Paper & Découpage	✓
Drag & Flog	☒
Freehand Paint	✓
Frottage	☒
Gild	☒
Grid Pattern	✓
Limewash	☒
Mask	✓
Monoprint	✓
Paint Mosaic	✓
Patinate	☒
Photograph & Projection	☒
Plaster	☒
Roller	☒
Splatter	☒
Stamp	✓
Stencil	✓
Sponge & Stipple	☒
Woodgrain	☒

THE PAINTS:
outdoor

Outdoor surfaces that need to be painted can pose a real problem, because not all paints are suitable for the outside. Demanding climactic conditions, which in a single day can vary from hot sun to pouring rain, will have a huge effect on how the paint will react and last in the long term. So, enormous effort is taken in testing these paints to ensure that they don't peel or wash away in a short time. We can truly say that these paints work the hardest. When you decide to paint outdoors, check the weather forecast and if it says rain, put off the job until another day.

There are many companies that specialize in outdoor paints and the hard-wearing paints they have created reflect the increasing interest in redesigning and decorating our gardens and exteriors, just like any other room in our home.

A surface such as concrete was, and still is, popular for paving stones and now there are paints available to change the colors of concrete patios. Bricks and cinder blocks can also be painted, so we can brighten up almost any surface. Admittedly, the color palette is limited, but we have choices now, where before we had none.

I think this is a very exciting time because pretty much any surface can be painted, from fences to plastic windows. If that old garden shed is offensive, you can camouflage it in green, blue, or silver birch. Go on, give it a dab of paint.

INSPIRATI

O N S

Wood Paint

This paint is the perfect answer for giving old garden furniture or decking a new lease on life. Apply it with a big brush and if you use a paint that contains preservative, there is no need to coat the wood first or varnish it afterward.

Concrete Paint

Concrete paving stones are an inexpensive paving material, but they really aren't very pretty. Paint a patio all one color, or if you are feeling adventurous, use several colors to create a checkerboard pattern. You apply the paint with a brush, so choose the biggest one you can find to make the job faster.

Brick Paint

This paint offers a great way of covering up cheap bricks or cinder blocks. Painting a whole house can take some time, as you put the paint on with a brush, but the end result can be brilliant.

Where?

Fabric	☒
Floors	☑
Furniture *seal absorbent surfaces*	☑
Walls	☑
Windows	☑

How?

Airbrush	☒
Colorwash	☒
Comb	☒
Crackle Paint	☒
Paper & Découpage	☒
Drag & Flog	☒
Freehand Paint	☒
Frottage	☒
Gild	☒
Grid Pattern	☑
Limewash	☒
Mask	☑
Monoprint	☒
Paint Mosaic	☑
Patinate	☒
Photograph & Projection	☒
Plaster	☒
Roller	☑
Splatter	☑
Stamp	☒
Stencil	☑
Sponge & Stipple	☑
Woodgrain *mix with glaze*	☑

Cork Paint

An inexpensive cork floor can be made to look much more glamorous with a coat of paint. This can be applied with a brush or short-pile roller, but it only works on unsealed tiles.

Melamine Paint

This paint and its associated primer can turn an unsightly kitchen into a contemporary one. The surfaces must be perfectly clean and free from grease before you paint and the best way to apply the paint is with a short-pile roller.

Plastic Paint

Make a splash by repainting faded plastic bathroom fittings or garden furniture. Easy to apply and available in funky colors, this paint is great. Some of these paints come as sprays and others are applied with a brush.

Radiator Paint

You have just finished decorating a room; the walls are perfect, the woodwork is smooth and glossy, the floor is neatly painted and varnished and the radiator is grubby and sad-looking. So paint it to tie in with the rest of the new color scheme.

Floor Paint

Sanded wooden boards can look great, but if they are badly stained, don't despair, you can paint them. This paint also works on concrete floors and most floor paints can be applied with a roller.

Glass Paint

Bathroom windows, glass kitchen cupboard doors and plain vases can all benefit from a little paint. The best way to apply it is with a brush and water-based paints should be sealed with varnish to protect them.

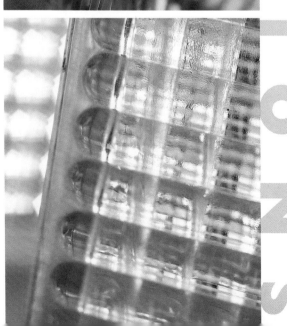

Not necessarily the most captivating paints in the world but still very important and not to be missed. These paints are for tricky surfaces in the home, which, once a room has been decorated, will standout as shabby, so it is worth reading on.

I like these special paints because they give previously unavailable options; a few years ago, we would have had to go to extreme lengths to change the look of our kitchen. Tired or out-dated melamine doors just had to be replaced. However, melamine paint has changed all that and now you can alter the color and look of your kitchen quickly and economically with a couple of pots of paint.

Tile paint in the past has posed a real dilemma, because the paint was not hard-wearing enough for day-to-day wear and so had to be treated gently. Now, rather than retiling a room, which is disruptive, hard work and expensive, use tile paint to get the job done. You can even use colored grouts to touch up between tiles after the paint has dried.

There is no doubt about it, radiators are boring to paint. I always use a short-pile roller and this does make the job much quicker. This paint is heat resistant and some manufacturers say you can also paint wood with it, so give the door a coat of paint to match while you're at it!

Using Specialist Paints

Where?

Fabric	☒
Floors	☑
Furniture	☑
seal absorbent surfaces	
Walls	☑
Windows	☑

How?

Airbrush	☑
use an air can	
Colorwash	☒
Comb	☑
mix with glaze	
Crackle Paint	☒
Paper & Découpage	☒
Drag & Flog	☒
Freehand paint	☑
Frottage	☒
Gild	☒
Grid Pattern	☑
Limewash	☒
Mask	☑
Monoprint	☒
Paint Mosaic	☑
Patinate	☒
Photograph & Projection	☒
Plaster	☒
Roller	☑
Splatter	☑
Stamp	☑
Stencil	☑
Sponge & Stipple	☑
Woodgrain	☑
mix with glaze	

There are literally dozens of different techniques you can use; whether you want a sophisticated finish or a bold texture, there is a technique that is just perfect. In this chapter, I have explored a whole range of traditional techniques and included lots of more contemporary ones that you may not have come across before. Each technique is demonstrated with easy-to-follow steps and a picture of the final effect. I have also included techniques for applying plaster, as I think that texture is important, and for other decorative finishes, such as gilding, which allow you to add all-important finishing touches.

None of these techniques require any special skills, though it is always a good idea to practice a new technique on a piece of board before you embark on a whole wall. Most of the techniques also

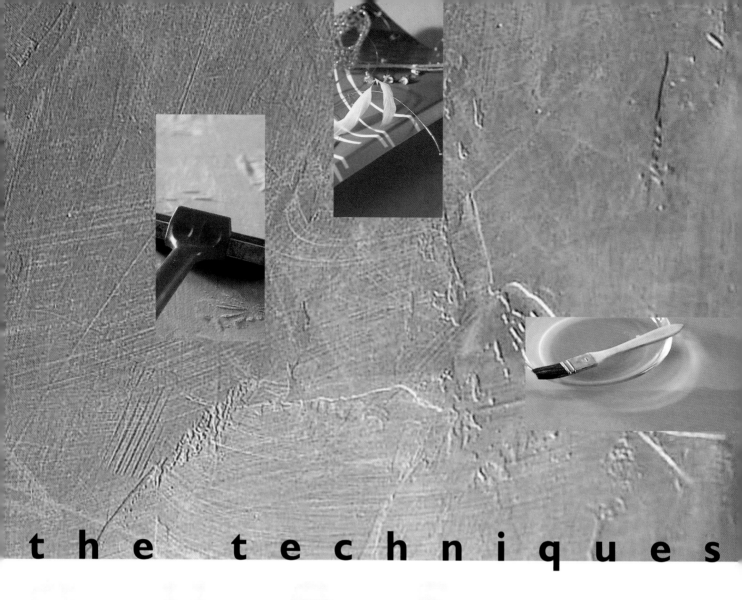

the techniques

require no special tools, just a roller or a paintbrush. I have used some specialized brushes to produce

particular effects and they are described in the relevant techniques. Generally, I use good quality

household paintbrushes: do buy the best you can afford, as cheap brushes will shed bristles, which can

be both fustrating to pick out and may spoil the finished effect. For glaze and colorizer techniques I

often use a varnish brush, which is thinner than an ordinary paintbrush and has a flatter end that gives

fewer brush marks. However, you can use an ordinary paintbrush instead.

When you have chosen a technique, refer back to the checklists in "The Paints" chapter to see

which paints are suitable. Prepare the surface, making sure that it is clean and smooth, and get started.

Aged Wood

If you have had to use nasty yellow pine in a room and you are desperate to cover it, this is the technique for you. This is a weekend project, as both the wax and the paint need to be left to dry at different stages. I think that this technique is best used on part of a wall—below a dado rail maybe—rather than all over it. Try dragging (see page 56) or airbrushing (see page 84) the top section of the wall for an interesting contrast.

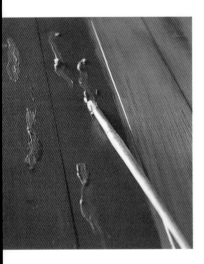

1. *Paint the wood with French enamel varnish, dilute one part varnish to two parts ethyl alchohol. When this is completely dry, paint the wood with emulsion in your chosen basecoat color. Leave this to dry completely. Using a small paintbrush, smear streaks of beeswax furniture polish down the wood, following the grain. Consider where the wood would get knocked or naturally wear, and that is where you apply the wax. Leave the wax to dry for at least 12 hours.*

2. *Dilute emulsion paint in your chosen topcoat color one part paint to three parts water. Carefully brush this mixture onto the wood, completely covering the surface and all of the wax. Leave it to dry.*

3. *Run a paint scraper over the surface, lifting off the smears of wax. Use a rag to wipe away any wax residue, leaving the wood looking softly worn.*

Verdigris Patina

This is both an easy and versatile technique to master. Use it on a vertical surface such as a wall or headboard, or on a curved surface like the large pot I have painted here. You can also verdigris wrought iron to give it an aged look. If you spray a plastic urn with enamel spray paint and then verdigris it, you can turn it into a stunning container for a conservatory. You can seal your work with spray varnish or diluted white (school) glue.

1. *Undercoat the surface, ideally with either bronze or gold paint and leave it to dry. Take three colors of emulsion paint—deep blue-green, pale blue, and mint green—and dilute a little of each with water, to the consistency of cream. Using a separate paintbrush for each color of paint and starting with the deep blue-green, trickle paint down the surface. Load the brush with paint, touch the bristles to the top of the surface, and just let the paint run down.*

2. *Continue to trickle paint down the surface, building up layers of the different colors. When you have a good mix of colors, covering most of the gold base, lightly spray spots of yellow ocher paint randomly over the surface. Then continue to trickle paint down the surface, over the spray until you are happy with the look.*

Sand Patina

As this technique is a bit messy—though it is also very satisfying—protect your floor by taping plastic sheeting to the skirting board before you start. This is quite a physical technique and you have to work quickly, in small sections at a time so that the paint doesn't dry before you can sand it.

Roll paint onto a one square yard section of the surface. Immediately sprinkle sand onto the surface, or, if you are painting a wall, take a handful of sand in your cupped hand and smear it onto the wet paint. With a household paintbrush, scrub the sand into the paint, coating the sand with paint and making brushstroke patterns on the surface. Leave it to dry.

Rough Plaster

This is the perfect technique for covering uneven walls, as it hides a multitude of sins. It is an easy and quick technique to master, and I find it very satisfying to do. If you can spread cream cheese on bread, you can do this.

Roughly trowel the plaster on to the surface, working on an area of about a square yard at a time. You don't want it to be too thick or uneven. Then just smooth over the plaster with a decorator's float to create the finish. Or, for an even rougher look, lay the face of the float against the plaster and pull it away to make tiny peaks. Then skim off the tips of the peaks with the float.

I plastered this wall with white rough plaster, left it to dry then painted it with colored emulsion paint (see page 10). I then mixed white emulsion with a little glaze (see page 20) and colorwashed (see page 58) the wall with it to really emphasize the roughness.

Imprinted Plaster

Before using this finish on a wall, experiment on a board to ensure the texture you are going to imprint shows. The fabric you use to imprint with will be ruined, so don't use a piece of antique lace. I used an old net curtain with an embroidered floral pattern. This technique is best used on a small area rather than a whole wall. Tint the plaster with colorizer (see page 20) before you start, as paint may obscure the imprints.

1. *Using a decorator's float, apply a thin, flat layer of plaster (tinted with colorizer if you want) to the surface. While the plaster is wet, lay the fabric on top of it and roll over it with a hard line roller.*

2. *Immediately peel the fabric off the plaster and leave it to dry.*

Stenciled Plaster

It is much better to use gesso (a kind of very fine plaster) rather than ordinary decorating plaster for stenciling, as gesso is far more manageable.

When you buy gesso it is solid and you must soften it in a double boiler until it reaches a workable consistency. Use an acetate stencil (see page 45) that can be washed easily when you have finished.

1. *Tape the stencil to the surface and then using a wide palette knife, working from the bottom up, smooth a thin layer of gesso over the whole stencil. The gesso should be the same thickness as the acetate stencil.*

2. *Immediately peel the stencil off the surface, working carefully to avoid smudging the gesso. Wipe the front and back of the stencil before using it again.*

Pearl-finish Plaster

This pink polished plaster wall was colorwashed (see page 58) with pearlized paint (see page 14) to enhance the polished look.

Polished Plaster

This is a similar technique to rough plaster, but the surface is worked to give a smoother, polished look. It is great for covering uneven walls and gives a more sophisticated look than rough plaster. It does require some energy, so make sure you are feeling fit before you start. If you want to color the plaster, as I did here, choose a deep base plaster, which has less white pigment, and add colorizer or acrylic pigment to color it. Mix the colorizer in with a whisk to ensure an even color.

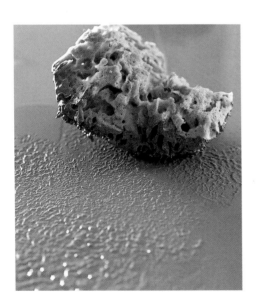

1 *Trowel the plaster onto the surface in an even, fine layer approximately $^1/_8$ in. (3 mm) thick. Then dab the plaster with a natural sponge to texture the surface.*

2 *Very gently skim over the surface with a decorator's float. When the plaster is dry, polish the surface further by rubbing the float over it, using a circular motion.*

Multicolored Horizontal Stripes

This simple technique uses different-sized rollers with colors that contrast and complement one another to create a striking horizontal arrangement. The timeless look has a retro feel with modern overtones. Choose a mix of bold colors, like those used here, or try the same technique using more neutral colors.

Roller Squares

Painting imperfect squares is an excellent way of breaking up a surface in an interesting way, and this roller technique is quick and easy to execute. These rollered squares have an uneven look that gives a personal touch. The combination of cream and silver used here make the design especially stylish and the colors work beautifully with the light.

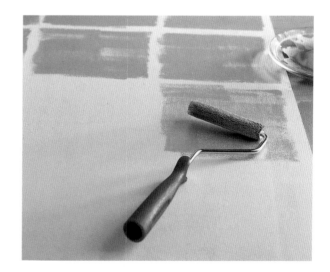

Decide on a pattern and colors by experimenting on a piece of board. Use a level and metal ruler to mark pencil guidelines horizontally across the surface. Pour different-colored paints into separate roller trays and choose a different-sized roller for each colored stripe. Roll on the paints, following your guidelines.

Measure and lightly mark out a square grid on the surface, based on the size of your roller and allowing for a gap of at least 1/4 in.(6 mm) between each square. Load the roller and use either a top-to-bottom or left-to-right stroke to fill in each square.

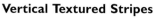

Vertical Textured Stripes

Give roller stripes some texture by winding masking tape around a roller in places. A good tip is to start with a punchy basecoat and use a paler, slightly translucent color for the stripes. For this particular wall, I have chosen to use the textured stripe quite sparingly, so that quite a lot of base color remains visible, but it would be just as effective to repeat the stripe more times.

Store-bought Roller

There are lots of different patterned and textured rollers on the market, so go and have a look in decorating shops to see what is available. These rollers are a great way of covering a large space quickly and can help make an expensive paint go further. For the topcoat for this sample, I used a metallic paint and to help it go further and give a more translucent finish, I diluted it half and half with glaze.

Wind masking tape around the roller at uneven intervals. Test the textured roller on paper to make sure you are happy with the results. Use a plumbline and metal ruler to mark vertical guidelines in pencil on the wall, then roll on the paint, starting at the top and working down.

Basecoat the surface with emulsion paint and leave it to dry. Roll the topcoat onto the surface as normal, or following any instructions on the roller itself.

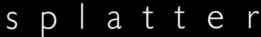

Small Splatters

Try this technique on a floor as I did, but your floor must be reasonably level or the paint will run everywhere. It doesn't work that well on floorboards, but if you have taken up carpet and underneath is hardboard, then it will work brilliantly. Again, it is messy so prepare the area before you start. Begin splattering in the furthest corner from the door, so that you don't paint yourself in, and leave the floor to dry overnight before walking on it.

Bleached Splatters

This technique is used on colored tissue paper. If you want to work it on white paper, use cold-water fabric dyes (see page 26) instead of bleach and splatter onto layout paper, which you can then cover a surface with (see page 70). Wear safety glasses when splattering with bleach, and wash your hands afterward. Also, bleach will mark your clothes, so wear old ones.

Dilute household bleach half and half with water. Lay the tissue paper out on a protected surface. Dip an old toothbrush into the bleach then, holding the brush in your right hand and a chopstick in your left, drag the bristles of the brush over the chopstick, directing the splatters at the paper. Leave the paper to dry flat on the surface, without moving it at all, and don't worry if it crumples a little where the bleach dampened it.

Basecoat the floor in your chosen color of emulsion paint and leave it to dry. Choose two colors of emulsion for the splatters and dilute each with water to the consistency of single cream. Mix up as much or more than you need to do the whole floor. Using a separate paintbrush for each paint, and working in an area of about one square meter at a time, dip each brush into the paint and lightly sprinkle paint of both colors over the floor. Keep the sprinkling fairly light for the best effect. When it is all dry, varnish the floor with two or three coats of tough, clear floor varnish.

Flicked Splatters

I really like this technique as it is a brilliant way of covering a wall quickly. The effect will run from about your height down the wall and as there is so much energy in the splatters, I like to use quite muted tones, as vibrant colors can be too strong. A word of warning; this is messy. Mask off everything and cover your furniture in plastic.

Basecoat the wall with emulsion paint and leave it to dry. Dilute the emulsion for the splatters to the consistency of single cream. Stand about 3 ft.(1 m) from the wall you want to splatter, dip a household paintbrush into the paint and, using your whole arm, flick the paint at the wall. Work along the wall, until you have covered the whole area. Be careful not to send the paint backward over your shoulder when getting ready to flick.

Gilded Splatters

Use this technique on a feature wall or a piece of furniture or glass (see page 64). It looks beautiful with light glancing off it and works especially well near floor uplights or accent lighting. In time, the edges of the splatters will start to tarnish, changing to a soft pinkish color. You could combine this technique with the flicked splatters technique and make one big gilded splatter across a wall.

1. *Using a soft-bristled artist's brush—a Japanese paintbrush is ideal—flick acrylic size over the surface. Leave it to dry until it becomes clear and tacky. Lay a sheet of metal transfer leaf face down onto a splattered area and rub the back of it so that it makes contact with the size. Peel off the backing and repeat the process until all the splatters are covered.*

2. *With a soft, thick brush—a clean blusher brush is perfect—gently brush away the excess leaf to reveal the splatters. The leaf clings to all the size, so every detail will show.*

43

Cardboard Stencil

This is the traditional method of stenciling, but it does have some limitations. It is best used with quite bold designs and you can use it only on flat or almost flat surfaces. You must use a special stencil cardboard— available from art supply shops—which you cut with a craft knife on a cutting mat. When you roll on the paint, do not load too much paint onto the roll; you want it to be half-dry so that excess paint does not run under the stencil.

1. *Draw the design you want to stencil onto the card with a permanent marker. If you want to trace a design, tape it to a window, tape the stencil card over the top of it and draw over the outline of the design onto the card. Lay the card on a cutting mat and cut out the design with a craft knife.*

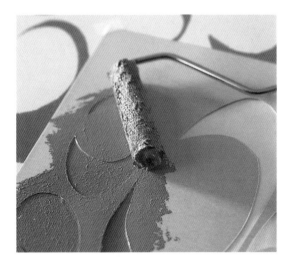

2. *Lay the stencil in position on your surface, holding it in place with stencil mount spray or masking tape if necessary. Roll over the stencil with your chosen paint—I used emulsion. Carefully remove the stencil and reposition it, making sure you don't lay it down on any wet paint.*

Acetate Stencil

Acetate is a much better medium for doing detailed stencils. The hot cutter goes through the acetate easily and lets you cut out intricate designs. If you are stenciling a small design then you could use a stippling brush to put the paint on, but I always use a roller as I find it quicker and easier.

1. *Lay your design on a work surface, lay the acetate over it and slowly trace over the lines of the designs with the hot cutter. The cutter will not mark the paper but will cut neatly through the acetate.*

2. *Lay the stencil in position, taping it in place if necessary, and roll paint over it. Remember not to load too much paint onto the roller. Reposition the stencil and repeat as needed. I angled the design a little on each repeat to give a more informal feel.*

Stenciled Fabric

You need to use pigment paint or thick hobby fabric paint (see page 26) to stencil onto fabric. This is an excellent way of giving new life to plain fabric, adding motifs to match other items in the room or adding accents of color. You can use either stencil card or acetate to stencil through.

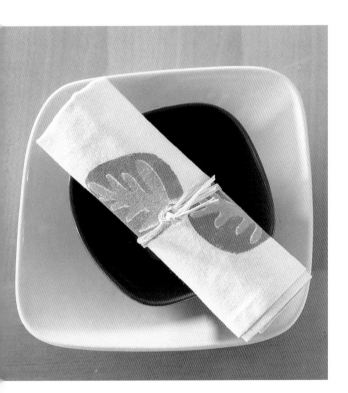

Lay the stencil in position on the fabric. With the tips of the bristles of a stenciling brush, or other fairly stiff brush, stipple fabric paint through the stencil. Follow any instructions on the paint and iron it on the back to fix it.

Stenciled Metal Flakes

These metal flakes are best used on walls or surfaces that are not touched too often. They are not suitable for furniture as they will get rubbed off. However, you can use them on the back of a glass tabletop, as I have here; the glass itself will protect the flakes (see page 64). I sprayed over the back of the stenciled design with spray paint (see page 18) and sealed it with varnish. Use a card or acetate stencil, depending on how intricate your design is.

Stenciled Curves

As neither stencil card nor acetate work very well on curved surfaces, the best way of stenciling a design onto a curve is with sticky-backed plastic. The plastic is low-tack, so it won't damage a painted undercoat. If you are working on a glazed surface then you need to prime it first. You can also use this technique on glass with frosting varnish to create a design of clear shapes on a frosted background (see page 18).

1. Lay the stencil in position and tape it in place. Roll acrylic size over the stencil then peel the stencil off. Leave the size to dry until it is clear and tacky.

2. Shake the flakes over the size, making sure that it is all covered. Gently rub the flakes into the size and pat them down so that they make firm contact. Brush away any excess flakes with a soft brush—a blusher brush is ideal.

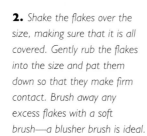

1. Basecoat the surface in your chosen color and leave it to dry completely. Cut out the design you want in sticky-backed plastic and stick it to the surface. Make sure that all the edges are well stuck down. Roll over the whole surface with your chosen topcoat. Leave it to dry, then roll on another coat and leave that to dry.

2. Use a craft knife to lift the edges of the sticky-backed plastic shapes and then peel the plastic off. Roll on one coat of acrylic varnish to seal your work.

Rubber Stamps

This is just a joy to do. It is so simple and so quick and you can buy everything you need in one stop at a home improvement store or specialized stamp shop. If you use the appropriate paint you can stamp onto literally anything; you can cover big areas or stamp just a tiny detail.

Use a small foam roller to roll paint onto the stamp. Press the stamp onto the surface. That's it.

Potato Prints

Revisit your childhood with this technique. In fact, children might love to help you decorate their rooms with potato stamps that they have designed and made (though small children should not be allowed to cut out their stamps themselves). Stamp with emulsion paint, then, if they make a mistake, you can just wipe it off.

Cut a large potato in half. Draw a simple design onto the flat front with permanent pen. Cut down around the edge of the design with a craft knife and then carefully cut in from the sides, removing the excess potato. Roll paint onto the raised design with a soft foam roller and stamp onto the surface.

Leaf Stamps

What is lovely about these leaves is that as they are so thin and the vein structure is so marked, they translate beautifully into paint. You can try the same technique with ferns or another strong botanical shape.

1. Lay the leaf on a sheet of paper and brush emulsion paint onto it.

2. Then lay the leaf on the surface and lay a clean piece of paper on top of it. Rub through the paper, rubbing your finger along each vein of the leaf toward the central stem. Peel off the paper and then peel off the leaf.

Sponge Stamps

You can use a cut out shape or a simple square, as I have here, to stamp with. I have used a synthetic sponge with large holes, to give lots of texture, and fabric paint (see page 26) to stamp this checkered tablecloth. You could also stamp size onto a wall and then gild it (see page 64); this would look particularly good over a fireplace mantle.

Dab the sponge into the paint, being careful not to overload it or the stamped design will not show any of the texture of the sponge. Stamp onto the surface three or four times, then dab the sponge into the paint again.

Rubber Comb

For all combing techniques (apart from combing onto fabric) you need to add glaze (see page 20) to emulsion paint to make it the right consistency. The ratio should be about half paint and half glaze. Mix or whisk them together until they are thoroughly combined. I chose a comb with teeth of decreasing sizes with this sample as it gives a pleasing effect with little effort. Combing works well on small areas such as below a chair rail. But you do have to comb in one smooth stroke and you cannot successfully join lines of combing in the middle, so it isn't easy to do a whole wall.

Metal Comb

Again, you need a glaze-emulsion mix for this technique. I have used the same comb I used for the oak finish (see page 68). If you are basecoating the surface as I have here, you must leave the basecoat to dry overnight before combing the topcoat or the metal teeth of the comb will damage the basecoat.

Paint on the topcoat and immediately comb through it. At the end of each comb stroke, wipe any excess paint off the comb (I find that my pants are perfect for this).

Paint the emulsion-glaze onto the surface using vertical brushstrokes. Drag the comb down through the paint, following the brushstrokes, in one smooth continuous stroke. Butt the edge of the comb up against the edge of the combed line and comb a second line. Repeat this until you have covered the whole surface. If you make a mistake, simply brush it out with the paintbrush. You don't need to put on more paint, just smooth out the mistake.

Combed Dragging

The contrast between the vertical dragging (see page 56) and horizontal combing is what makes this sample so interesting. The moiré effect where the combed lines cross is wonderfully three-dimensional. You could make more comb lines, just leaving tiny areas of dragging, for a less obvious, more intense, look.

Drag the surface with your undercoat color as described in one-way dragging. Leave this to dry completely. Paint on the topcoat of emulsion-glaze and immediately comb through it with a rubber comb, using sweeping strokes that intersect at intervals. The dragged background will show through the emulsion-glaze in the uncombed areas.

Combed Monoprint

You cannot comb straight onto fabric, it just doesn't work. You have to do it via monoprinting (see page 86). The amount of paint on the acetate is crucial; too much and the finished design will be blobby; too little and the design won't show properly. Experiment on a scrap piece of fabric before you work on the main piece.

1. Brush undiluted fabric paint onto a sheet of acetate and comb through it with a plastic comb.

2. Lay the acetate paint-side down onto the fabric and smooth over the back of it, up and down the combed lines, with your hand to make sure that all the paint has made contact with the fabric. Peel off the acetate. Iron the fabric on the back to fix the paint.

Homemade Comb

While there are a lot of combs commercially available on the market, sometimes you just can't find the right one. So, make one. You need a stiff card that is not too thick—the cardboard back of a notepad is ideal. For this sample, I painted on the topcoat with neat brushstrokes and then took the comb through the emulsion-glaze in sweeping strokes.

Draw the comb you want onto the card and then cut it out with scissors. I snipped off the very tips of the points of this comb to make the lines a little thicker. Paint on a topcoat and comb as normal.

Corrugated Cardboard Comb

This technique works well if you use a basecoat and topcoat of contrasting tones of a color. One of the really pleasing things about this is that the repeat is invisible across the wall; you can't see where the new stroke starts.

Cut across the corrugations in the cardboard to give a combing edge with the corrugations showing. Paint on the topcoat and immediately drag the card across it. I have allowed the strokes to wobble a little to enhance the random effect.

Pictorial Frottage

Use this technique to add an appealing, feminine element to a painted wall. It works very well in a bedroom or a hall and you can work on a large scale, which will add drama to the room. Don't panic if you can't draw; use the projection technique (see page 72) to create your image and it will look wonderful. Paint it in ordinary emulsion paint and use newspaper or layout paper to do the frottaging.

1. *Draw out the whole design on the surface in very faint pencil lines. Paint the image a small section at a time. If you are worried about applying color in the right areas, you can project the image onto the wall and then leave the projector on while you paint; this makes it very like painting-by-numbers.*

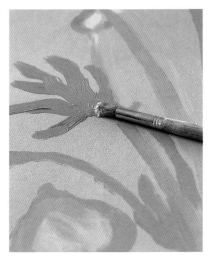

2. *With your hand, smooth a piece of paper over the painted area, making sure that all the paper makes contact with the paint, and immediately peel the paper off again.*

Abstract Frottage

I do love this technique; you can add to it, build up on it, and just paint over mistakes. There really are no rights or wrongs. You do need to be a bit organized before you start; tear up strips of paper to frottage with, making sure you have enough for the whole wall. You need to work quite quickly or the paint will dry. To create gently swooping guidelines, I just walked along the undercoated wall holding a pencil against it at different heights.

Undercoat the wall in a medium tone so that you can add frottaged stripes as highlights. Paint on a stripe, which can be up to four times the length of a piece of your paper. Immediately smooth the paper over the paint, a section at a time, then peel it off. Frottage along a whole stripe before painting the next one.

Rolled and Frottaged Squares

These roller squares have been softened and given a two-tone texture using the frottaging technique. Use a basecoat similar in color and tone to the topcoat for a subtle effect, or choose contrasting colors for a more textured look.

Undercoat the surface and leave it to dry. Lightly mark out a rough grid, based on the size of your roller and allowing for a gap of at least ¼ in.(6 mm) between each square. Roll paint onto one square. Smooth a sheet of paper over the paint. Peel off the newspaper and repeat on subsequent squares.

One-way Drag

There are a lot of ready-mixed paints you can drag with, but it is also simple to make your own. You cannot use ordinary emulsion, as it will dry before you can finish the dragged effect. It is also too thick; the background color won't show through. For the wall below I made a dragging paint by adding colorizer (you can also use acrylic) to scumble glaze (see page 20). Add colorizer until you reach the shade you want. Before you start, mask off areas you don't want to paint, as you have to do this technique quite quickly and you won't have time to cut in next to a ceiling or architrave. It is also better to work with someone; one of you rolls the paint on and the other drags it.

Undercoat the surface with emulsion in the background color and leave it to dry overnight. Roll the tinted scumble over the painted surface, working in yard-wide strips at a time. Then, working from top to bottom, and as far as possible using one smooth stroke, pull a wide dragging brush across the surface. If one stroke is impossible, drag from the top downward, then from the bottom upward, feathering out the brushstrokes at the ends so that they meet gently in the middle.

Two-way Drag

This technique is worked in a similar way to one-way dragging, but for the wall above, I used a different paint recipe. I chose a colored emulsion and then mixed a glaze into it (see page 20). The proportions are approximately one-third emulsion to two-thirds glaze. This recipe will dry more quickly than the scumble recipe, but you can use your favorite colored paint. If you are doing this in hot weather, use a special tropical glaze that doesn't dry so quickly.

Flogged Finish

This technique also needs a special paint and you can use either of the recipes given for dragging. This is quite a delicate finish, but you can make it a little punchier by using contrasting colors for the undercoat and topcoat. If you are using this technique to cover a whole wall, work in yard-wide strips and do bear in mind that, though simple to do, this technique is quite labor-intensive.

Undercoat the surface and leave it to dry. Roll on the emulsion-glaze or tinted scumble. Using a dry, long-bristled flogging brush, literally flog the wet surface; you should do this with a wrist action rather than using your whole arm.

Undercoat the surface and drag it as described in one-way dragging. Leave it to dry and then roll over the dragged surface with more dragging paint and drag it again, this time working at right angles to your first drag. So, if the first drag ran from top to bottom, the second drag should run from left to right.

One-color Colorwash

This is such a simple technique and although it has been around for a while, it is just as valid now as it ever was. This is a technique that is best done with a partner—one of you paints and the other wipes.

1. *Basecoat the surface in a plain emulsion and leave it to dry. Mix one part of emulsion paint with three parts of glaze for the topcoat. Working in a square meter area at a time, use a wide paintbrush to paint the topcoat onto the surface. Decide on the direction you want the colorwash to run on the surface and apply the brushstrokes in that direction.*

2. *While the paint is still wet, gently wipe over it with a rag, following the direction of the brushstrokes. Use sweeping strokes to create an expressive surface.*

Two-color Colorwash

To develop the basic colorwashing technique further, build up the layers, using toning shades for each layer. The basic principle is the same, but for this treatment pour some scumble glaze into a bowl and add artist's acrylic color until you achieve the shade you want. This gives a more translucent paint that allows the first layer of colorwashing to show through.

Colorwash the surface as described in one-color colorwashing and leave it to dry overnight. Mix a toning color of glaze and repeat the process.

Colorwashed Wood

Here is a variation on the colorwashing technique that can be used on floorboards. Rather than diluting paint, it is best to use a product specifically designed for the purpose.

Apply a coat of the color and immediately wipe off any excess with a rag to give a soft colorwashed finish. You can use up to three coats of color, wiping each down in between, depending on the depth of the shade you want.

Colorwashed Stripes

Colorwashing doesn't have to be used over a whole wall. Try masking off stripes (see page 76) or squares and colorwashing only them.

Basecoat the surface and leave it to dry. Mask off the areas to be colorwashed. Mix one part of pearl paint to three parts of glaze and colorwash the masked areas. Immediately and carefully peel-off the tape.

You can add another dimension to a colorwashed floor by stenciling a design onto it in woodstain or floor paint. Seal the finished work with two or three coats of varnish.

1. Rub a wire brush over the wood, following the grain, to open it up a little. This will allow the paint to be absorbed more easily.

2. Dilute two parts of white emulsion paint with one part of water. Paint the wood using a large paintbrush and following the grain. Leave it to dry for a few minutes then wipe off any excess paint with a rag.

Painted Limewash

For a quick and easy-to-do limewash effect, use paint. The finished look is very similar to traditional waxing and if you seal it with varnish, it will be just as durable.

1. Basecoat the surface with oil-based eggshell paint and leave it to dry completely. I chose a deeper color than my topcoat to show off the contrast to the best effect. Using a large paintbrush, brush a topcoat of emulsion paint over the eggshell. Keep the brush quite dry and use random strokes to apply the paint. The basecoat should be covered, but not completely—the color should show through the strokes a little. Leave it to dry completely.

Waxed Limewash

This is the traditional method of limewashing wood. You need to be feeling energetic if you want to do this over a large area as it requires a lot of elbow power and does take some time.

1. *Rub a wire brush over the wood, following the grain, to open it up ready for the wax.*

2. *Dab a rag into white liming wax and rub it into the wood using a circular motion. Work on a small area at a time, rubbing the wax well in and then using a clean rag to wipe away any excess wax.*

Scrubbed Limewash

You can use this paint effect on flat walls rather than real wood to simulate a limewashed effect. For a total look, use this paint finish in a room with a limewashed floor.

2 *Saturate an old rag in denatured alcohol and gently wipe away areas of the topcoat, exposing small patches of the basecoat.*

Cardboard Printed Grid

This is good fun and you can make the grid as irregular or regular as you like; the broken lines give it an informal feel that works either way. Try doing one wall in a room with small squares and another wall in exactly the same way, but with huge squares.

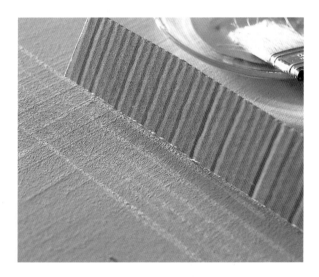

Cut out a strip of corrugated cardboard, cutting across the corrugations. The cut edge needs to be straight so that all of it makes contact with the paint and the surface. Use the bristles of a household paintbrush to dab paint onto the edge of the card and then press the edge against the surface.

Taped Grid

This technique offers a great way of copying a tartan design, either in traditional colors, or you can give the pattern a contemporary twist by painting it in unusual shades. The trick is to spend time measuring and masking the pattern off. It is best to mask (see page 76) and paint all the horizontal lines first, leave them to dry and then do the vertical lines.

1. *Basecoat the surface in emulsion paint. Decide on the pattern you want to paint and then measure it out on the surface, marking the ends of the lines with small pencil marks at the top and bottom and each side of the surface. Press low-tack tape onto the mark at one side of the surface and stretch it across to the corresponding mark on the other side. Press it down along its length. When you have masked all the horizontal lines, paint them in with a brush or small roller and emulsion paint.*

2. *When you have painted all the lines, carefully peel the tape off. Remember that it will have wet paint on it, so ensure that you don't get any on the wall. You can leave the tape to dry and use it again for the vertical lines, which you do in exactly the same way.*

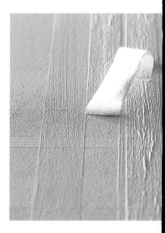

Simple Roller Grid

For a really simple, informal grid pattern, just use a little foam roller. As it is quite quick to mark out and do, this is an ideal technique for a large area. Also, the sponge roller will help make the paint go a long way, so you can afford to use an expensive pearl or metallic finish (see pages 14 and 16).

Taped Roller Grid

It is easy to customize a roller to make a grid. Choose a medium-pile roller and wind narrow masking tape around it at intervals. It gives a less uniform effect if the gaps between the tape are of different sizes, though you must then remember always to roll with the handle on the same side, or the lines won't match up.

Mark out the pattern on the basecoated surface with pencil dots where the lines will cross; part of the charm is the slight irregularities in the lines. Roll the roller through the paint and roll the horizontal lines first, then the vertical ones. If you need to recoat the roller with paint, stop at a point where the lines intersect so the join will show less.

To make a grid, mark out the central line of each horizontal and vertical section of stripes with a chalk line on the basecoated surface. Chalk line kits are available from home improvement stores and come with instructions. Roll the horizontal lines, then the vertical ones, with the central section of the roller following the chalk lines.

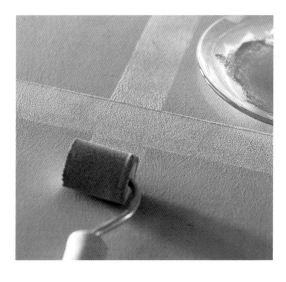

Gilded Glass

Gilding is one of those techniques that looks so impressive that you think it must be difficult to do, but it really isn't. There are various colors and types of transfer metal leaf available; gold, Dutch metal (which looks like gold), patinated Dutch metal, silver, aluminum (which looks like silver), and copper. Choose whichever works best with your color scheme. This piece of glass was gilded with 22-karat gold leaf and the frame is rolled with metallic paint (see page 16). The gilding is done on the back of glass, so you will see the finished effect through the glass itself.

2. *One sheet at a time, lay the gold leaf face down on the sticky size. Rub the back of the paper with your hand to make sure that all of the leaf has come into contact with the size. Peel off the paper backing. Butt the sheets of leaf up against one another, but don't worry about small gaps as they can either be filled in later or, if you are going to distress the leaf (as I will here), they will add to the aged look.*

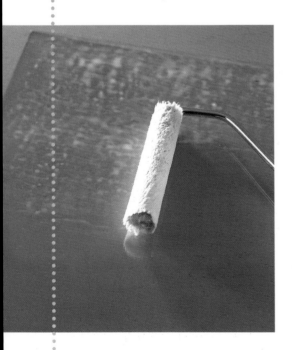

1. *Using a short-pile roller, roll acrylic size onto the back of the glass. Leave this to dry until it becomes clear and tacky; you can test it with the tip of your finger to see if it is ready.*

3. *Using fine wire wool, gently rub the leaf. For an evenly distressed look, progress across the glass, working in smooth strokes from side to side. For a more random look, rub the wire wool in small circles across the leaf. Either way, you cannot tell from this side how much leaf you are taking off; you need to keep turning the glass over and checking the front to see what the effect looks like.*

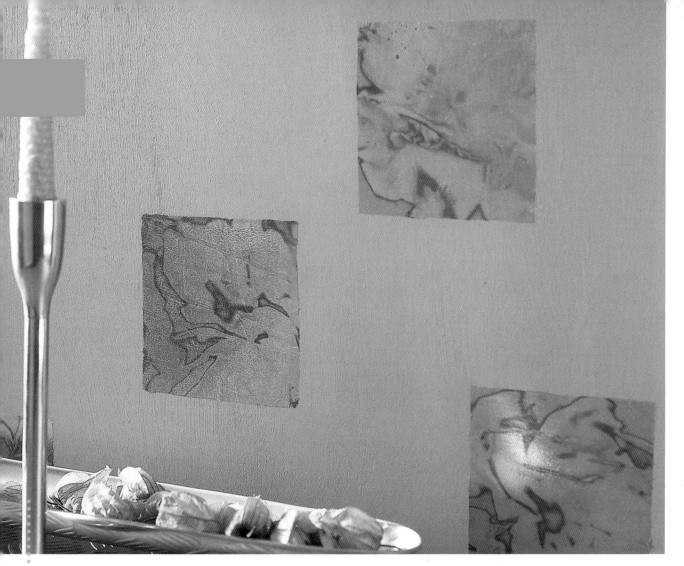

Gilded Patchwork

As I was only treating a small area with this technique, I drew the squares onto the surface and then painted each one with acrylic size using a small paintbrush. If you were covering a large area, however, this would take ages. Therefore, I suggest that for a large area you cut a card stencil (see page 44) to the right shape and roller the size onto the surface through it. I have used a patinated Dutch metal leaf, which is available in various colors, on this sample.

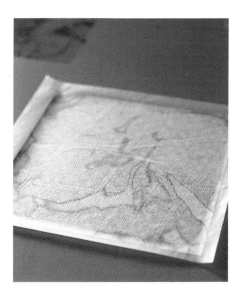

1. *Mark out the shape you want to gild onto the surface and paint it in with acrylic size. I have chosen a simple square, but you can paint any shape you like. Leave the size to dry until it is clear and tacky.*

2. *As before, lay the sheet of metal leaf face down on the size and rub the paper backing before peeling it off.*

65

Gilded Stamps

This is another way of applying transfer leaf to a surface. It is similar to the stamping with leaves technique (see page 49), but allows you to use the brilliance of metal leaf. You could stamp with a variety of things, but do experiment on a board before you start on a whole wall.

Gilded MDF

MDF is a good surface for gilding onto as it is so flat. The metal leaf is amazingly thin and will hug any tiny flaw in the surface, so do make sure that there are no irregularities in the surface before you start. You can gild onto wood, but all the grain will show through. I gilded this frame in 22-karat gold, which I find easier to use than Dutch metal, although it is more expensive. If it is within reach of your finances, I do recommend 22-karat gold.

Roll the size onto the MDF and leave it to dry until it is clear and tacky. Lay a sheet of gold leaf face down on the surface and rub the paper backing. Peel off the backing and place the next sheet. Continue until you have covered the whole surface. If you want to put the item in a bathroom, seal it with a special varnish, available with the leaf. You can paint the varnish on with a brush or spray it with an air-gun (see page 84).

2. *Position the skeleton leaf on the surface and then lay a clean sheet of paper over the back of it. Rub the leaf against the surface, through the paper. Rub your fingers along the veins toward the central stem. Peel the paper and leaf off the surface. Leave the size to dry until it is clear and tacky.*

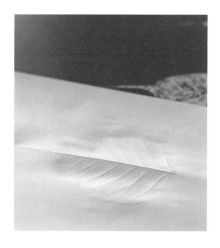

1. *Lay a skeleton leaf on a clean sheet of paper and paint the surface with acrylic size. Depending on how much size you apply and how successfully the size is transferred to the surface with each stamp, each gilded leaf will look slightly different.*

3. *Lay a sheet of transfer leaf—I have used silver—face down on the size and rub the paper backing, then peel it off. With a thick, soft brush—a clean blusher brush is ideal— brush away any excess metal leaf to reveal the detail of the veins and texture of the skeleton leaf. The silver leaves will tarnish in time, but you can roll a thin layer of acrylic varnish over the whole surface to seal it.*

Maple

This is a rather labor-intensive effect and is probably best used on smaller areas like a tabletop or a panel. After all, you wouldn't really have a maple wall. You do need the right brushes to create this effect; a mottler brush and a softening brush. Make up a paint by mixing a scumble glaze and a colorizer together (see page 20). You can buy special wood colorizers to make your own warm, light brown glaze.

Oak

You can use this technique to cover a whole wall to wonderful effect. Use the same paint recipe as for the maple effect, but adding a light oak colorizer, or use a burned umber acrylic to tint the glaze. Undercoat the surface with a cream-colored emulsion. It is important that the undercoat is completely dry or the metal comb will damage it.

1. *Undercoat the surface with honey-colored emulsion and leave it to dry completely. Brush the glaze over the emulsion with a flat paintbrush.*

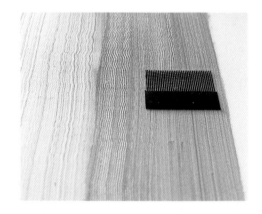

2. *With the mottler brush (above), make a narrow zigzag pattern in the glaze. Then, using the tips of the bristles of the softening brush, gently stipple over the zigzags, softening them a little.*

Brush on the paint in the direction you want the grain of the oak to run. Then drag a metal woodgraining comb through the paint, following the brushstrokes. As you drag the comb, move it a little from side to side, with a gentle rippling action, to simulate woodgrain. To enhance the effect, you can then sweep the comb randomly up and down the surface, breaking up the grain a little.

Knotted Wood

This is an ace technique that you can use to cover large or small surfaces. Again, mix scumble glaze (see page 20) with colorizer —I used a medium oak here—to make your own paint. A soft brown undercoat works well with this color glaze. You need a special woodgraining rock and roller (available from home improvement stores) to create this effect. There is a bit of a knack to this, but the glaze remains workable for some time, so if you make a mistake, just brush it out and start again.

Brush on the paint in the direction you want the grain of the wood to run. Starting with the rock and roller tilted away from you, run it down through the glaze, rocking it backward and forward to create the knots in the wood.

Contemporary Knotted Wood

This sample just goes to show that you can use vibrant colors to create amazing faux wood effects using a traditional technique. The undercoat of this sample was bright yellow emulsion and the topcoat of glaze was a strong pink.

Crushed Paper

You need layout paper for this technique. These large sheets of paper can be bought in pads from art stores. This paper is thin and can be collaged onto a surface without the joins showing. The paper is colored with water-based fabric dye (see page 26), so do wear gloves if you don't want to stain your hands. If you make up enough sheets of this paper, you can cover a whole wall with them. The finished paper looks very crumpled, but glue it to the surface with diluted white glue, smoothing it as you work and it will dry flat. Using a large sheepskin roller, seal it with a water-based matte varnish.

Salt-dyed Paper

Again, the color comes from cold-water fabric dyes (see page 26). You cannot move the paper once the salt has been sprinkled on until it is completely dry. So, be organized and lay down layers of newspaper to protect your work surface. You can cover a wall, or a piece of furniture, with this paper, gluing it on with diluted white glue. If you want a finish with a slight sheen, use a large sheepskin roller and water-based satin varnish to seal the wall.

Brush diluted fabric dye onto the paper and while it is wet, fold the paper in half, then, using your hands, crush it into a tight ball. Immediately flatten it out again and leave it to dry flat. The side that you applied the dye to will be the back of the finished sheet.

Mix up the fabric dye following the instructions to create a strong color. You can mix colors together to create your own shades. Using a big, soft brush, paint a sheet of paper with the dye. Immediately sprinkle salt onto the wet dye; you can use any kind of salt. Sprinkle it as though you were sowing seeds. The dye attracts the salt around it, leaving areas of a lighter color. When it is dry, take the paper outside and brush off the excess salt.

page

Découpaged Photocopies

I chose to use leaves because, as they are flat, I could lay them straight onto a color copier and photocopy them. If you want to use a three-dimensional object, like a flower, it would be best to photograph it and then color photocopy the photographs.

Simply cut out the leaves with scissors and arrange them on the item you are decorating. When you are happy with the arrangement, glue the leaves in place with white glue. Wipe away any excess with a damp cloth. When everything is dry, varnish the item with water-based varnish.

Découpaged Glass

This technique is done on the reverse of the glass so, alarmingly enough, you have to put the white glue onto the face of your painted paper. The glass protects your work perfectly. I used black and white photocopies as the basis for the découpage.

1. *Color the photocopies by painting them with either cold-water fabric dyes, inks, acrylics, or artist's watercolor paints. When they are dry, cut out the painted designs.*

2. *Brush white glue onto the front of each painted piece and lay it face down on the back of the glass. Wipe away any excess glue. When it is dry, the glue will be clear. You can then spray over the painted paper with colored or frosted spray (see page 18). When you turn the glass right-side up, the color or frosting will work as a background for the paper. You can also seal the spray with oil-based clear varnish.*

HOW TO CREATE THE EFFECT

1. *Brush diluted white glue onto the back of the photocopies. Lay the photocopies on the surface, smoothing them out flat. If you trap air bubbles, either smooth them out or prick them with a pin. Wipe away any excess glue.*

2. *Mix scumble glaze and colorizer (see page 20) to tone with the undercoat. When the photocopies are dry, colorwash (see page 58) over them, and any bare areas of wall.*

Painted Photocopies

This is an excellent way of adding faux friezes or panels to a painted wall. Arrange them randomly, as I have done here, or as a border. Once you have found an image to inspire you, just black and white photocopy it as many times as necessary and cut the photocopies out. Undercoat the surface with emulsion paint in your chosen color and leave it to dry before you start.

Projected Images

If you are not confident about drawing and painting an image onto a wall, this is the technique for you. It is the decorator's equivalent of painting-by-numbers. If you don't have an overhead projector, you can rent one. Make the image to be projected by photographing your chosen object and then color-copying the photograph onto acetate. Alternatively, use a slide projector and project a transparency.

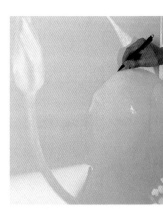

1. *Project the image onto the wall, moving the projector until it is in the right place and is the size you want. Leaving the projector on, draw around the image with a pencil.*

2. *If you want to copy the projected image exactly, then leave the projector running while you paint. If you are inventing your own colors or painting a silhouette, turn it off. Either way, simply paint in the outline with emulsion or artist's paints.*

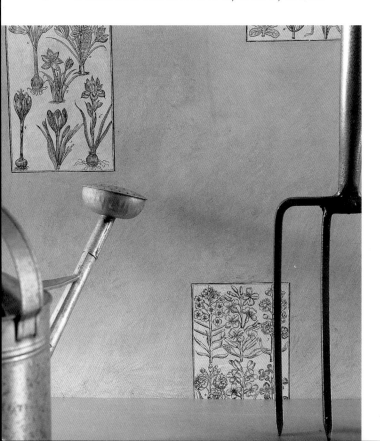

projection

Lasertran

You can transfer almost any image to almost any surface with this extraordinary product. Wood, wax, stone, plaster, or white can all be decorated with amazingly lifelike copies of whatever you want, from a flower to a piece of toast. A separate type of film allows you to transfer images onto fabric.

1. Cut out the design and soak it face-down in a bowl of water for approximately one minute.

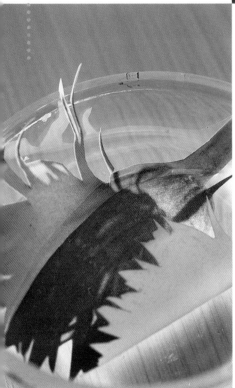

2. Take the design out of the water and lay it face-down on a sheet of newsprint to remove any excess water. Dip a rag in pure turpentine and wipe over the surface you are going to stick the design to.

3. Working quickly, lay the design face down on the turpentined area and slide off the backing paper. Dab it with the turpentined rag to make sure the design hugs any variations in the surface. Leave it to dry flat.

Natural Sponge

Sponging is a well-established technique, but just because it has been around for a while, you shouldn't ignore it. The colors that you use make a big difference here; if you use dated colors the technique will look terrible. Don't choose colors that contrast strongly; this technique will look more contemporary done in toning colors. Also, you need to make the finished surface look well-worked and quite densely covered for the best effect.

Basecoat the surface with emulsion paint and leave it to dry. Mix the topcoat emulsion half and half with glaze. Dip a natural sponge into the paint and then dab it onto the surface, working in square yard sections. Keep dipping and dabbing until you have achieved the right look then move on to the next section.

Synthetic Sponge

A synthetic sponge gives a much finer texture than a natural sponge. If you trim the sides of the sponge to make a ragged shape, you will avoid any hard edges. This technique works best on a completely flat surface; any small imperfections will show. I took my time with this sample to ensure that the surface was evenly and quite densely covered, which makes the finished effect look rather like leather. To complement the subtlety of this finish, I decorated the skirting board using the scraffito technique (see page 91).

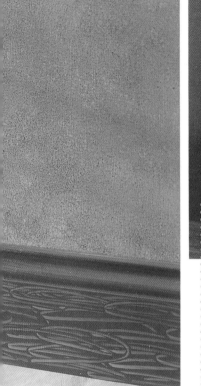

| e

Light to Dark Stipple

Stippling is quite a time-consuming technique, but the end result is great. It is a very subtle look and to make the most of it I have graduated the colors so that the fine texture shows up. If you are going to use this technique on a large wall, please do buy the largest stippling brush you can find, or the task will just be painful. You may find it easier to work with a partner; one person paints and the other stipples.

Basecoat the surface with emulsion and leave it to dry. Mix the topcoat emulsion half and half with glaze, dip the sponge into the paint and dab it onto the surface. As before, keep dipping and dabbing until you have achieved the right finish. Don't overload the sponge with paint or you will lose all the texture.

1. Basecoat the wall in the lightest color emulsion and leave it to dry. Mix the midtone paint half and half with glaze: if the weather is hot then use a tropical glaze that will give you more working time. Roughly measure up from the floor to the height you want the midtone color to reach to and make faint marks across the wall. Paint the midtone on to the surface up to just below the measured height. Working quickly, stipple the entire wet surface. When you reach the top, carry on stippling 2 in. (50mm) above the line to create the soft edge. Don't put more paint on the brush, just use the paint already on the bristles. Leave to dry overnight.

2. Repeat the process with a darker color on the bottom section of the wall. While stippling, you may find that the brush becomes clogged with paint, in which case just dab the excess off onto a rag.

Masking Stripes

I wanted to make these stripes irregular to make sure that they didn't look like wallpaper. This is quite time-consuming, as you have to tape some stripes, paint them, leave them to dry and then tape other adjacent stripes. Therefore, I think that this is a technique for a feature wall or alcove rather than a whole room.

Apply a basecoat to a wall in emulsion paint and leave it to dry. Use a low-tack tape to mask off stripes and then paint them in with different colored emulsions. Immediately peel off the tape. If you hang the tape up to dry, you can use it again on further stripes.

Masking Curves

For a more fluid look you can mask off curves or curved stripes with special curved tape or narrow straight tape. The advantage of masking stripes is that you get a neat edge, which you wouldn't if you were painting them freehand.

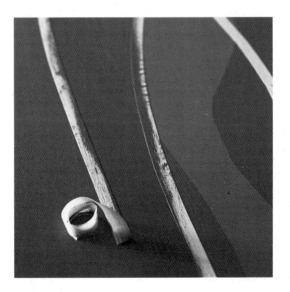

Use narrow masking tape that has some "give" in it to mask off gentle curves on an undercoated surface. Put little pleats in the outside edge of the tape to help it curve, but you must not pleat the inside edge as it will put "steps" into the painted edge. For tighter curves, use electrical tape. I mixed a little glaze into the topcoat emulsion and once I had painted the stripes, I pushed the bristles back through the paint to create some texture before peeling off the tape.

77

Masking Fluid

Masking fluid can be bought in art shops and looks like runny rubber adhesive. It can ruin your paintbrush, so don't use a good one; though you can scrub the bristles, from root to tip, with a wire brush to remove the fluid. This is a good technique for a small area as it takes time and effort to rub the fluid off. Try it when you feel energetic.

1. *Apply a basecoat to the surface in emulsion paint and leave it to dry overnight. Paint the fluid onto the surface using bold sweeps—get your whole arm moving to create an exuberant expression. Leave it to dry; it will become clear when it is ready.*

2. *Mix paint half and half with glaze (I used colored metallic paint) and paint or roll over the surface, covering the fluid and basecoat. Leave it to dry overnight.*

3. *Here is the time-consuming part. With your fingers, rub the surface where the dried fluid is. The fluid will loosen and peel away. It is important to remove every last scrap of fluid for the best effect. If it makes your fingers sore, wear a clean leather gardening glove.*

Masking with Wax

This technique is sometimes known as batik and is specifically for masking onto fabric. You need to buy a special wax, available from craft shops, and melt it in a pan over a very low heat on the stove. Don't leave the pan for a moment while you are melting the wax, as it can catch fire suddenly and silently.

1. Stretch a piece of fabric evenly over a wooden frame (you can make or buy one of these). Paint the design onto the fabric using a paintbrush and the melted wax. I painted a freehand design, but you could draw out the shapes first with a dressmaker's pencil.

2. Mix four parts of cold-water fabric dye with one part of baking soda. Then mix one teaspoon of this dye mixture with $^1/_4$–$^1/_3$ of a cup of warm water to produce a concentrated color. This dye will remain active for only 30

minutes, so don't mix up too much at a time. Paint the dye onto the fabric with a brush. The areas covered in wax will resist the dye. Leave the dye to dry on the fabric for as long as possible to get the strongest colors– up to a week is ideal.

To remove the wax, take the fabric off the frame and immerse it in boiling water. Stir it for one to three minutes then scoop off any wax that has risen to the surface. Rinse the fabric in cold water and flake off any excess wax with your fingers. Iron the fabric. If you are painting a large item you can have it dry-cleaned to remove the wax.

Stamped Mosaic

This is the best way to cover a wall in faux mosaic. It is simple and fairly quick to do. One of the nice things about it is the slight irregularities; some of the squares are fainter than others and nothing is perfectly straight. When you have finished, carefully wash the sponge stamp (see page 48) and save it for use again.

1. *Make a simple stamp from squares of synthetic sponge (I used a bread knife to cut them) and glue them onto thin MDF or hardboard. Use undiluted white (school) glue to attach the sponge to the MDF and let it dry before you use the stamp.*

2 *Basecoat the wall in emulsion paint and leave it to dry. Mix the topcoat emulsion half and half with glaze. Dab the sponge into the paint and press it onto the wall. Without recoating the stamp with paint, press it onto the wall again, lining it up with the first stamped section. Continue until the whole surface is covered, recoating the sponge with paint about every fourth imprint. Don't touch the surface until it is dry.*

Stenciled Mosaic

Use a stencil to create mosaic borders on walls and floors. I used acetate to stencil with (see page 44) as I had devised a fairly intricate design. I colorwashed the wall first (see page 58) and stenciled the border on once that was dry. You could stencil a border then use the stamped mosaic technique to fill in the central area.

Painted Mosaic

Hand-painting mosaic is a very time-consuming process, so I suggest that you use it only as a small detail. I used a flat, square-ended brush to make it easier to paint the squares. To make this plain vase more glamorous, I added this glittery mosaic band around the top.

Paint the mosaic tiles onto the vase with glitter gel ceramic paint and then sprinkle more glitter onto the paint while it is wet. When it is dry, dab a little more paint on top of the glitter to seal it.

Lay the stencil in position and tape it in place. Roll over it with emulsion paint. Lift off the stencil and reposition it and repeat the process.

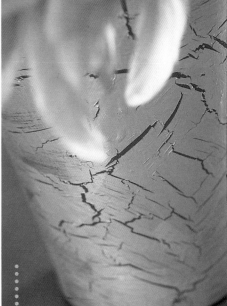

Metallic Crackle

This metallic crackle paint gives the most stunning effect over a basecoat of colored emulsion paint. Choose a more traditional antique look by using a basecoat of soft blue or red, or opt for an unusual, contemporary effect by applying a basecoat in a vivid color such as lime green or strong lilac.

Gum Arabic Crackle

This is a very effective technique for producing a crackled finish, but I advise you to practice on a board before starting a project as there is a knack to getting it right. For a contemporary version of a classic look, work the finish in vivid, rather than traditional soft pastels.

2. *Load-up a roller with emulsion paint and roll it once over the gum arabic. The trick here is to apply this topcoat as thickly as possible in one coat. You cannot go back and forth with the roller to apply more paint as this will ruin the effect. The cracks will appear very quickly and will enlarge as the paint dries.*

1. *Paint the item, in this case a terra cotta plant pot, with a coat of emulsion paint and leave it to dry. Then, paint the pot with crackle glaze and leave this to dry completely.*

2. *Using a soft brush, paint on the metallic topcoat. As with the gum arabic crackle finish, you can paint an area only once, so it is important to have a lot of paint on the brush. If you go back and put on more paint, you will spoil the finish. Again, the cracks will appear almost immediately and will get bigger as the paint dries.*

1. *Paint the surface with a basecoat of emulsion paint and leave it to dry. This color will show through the cracked surface. When the emulsion is dry, paint the surface with a fairly thick layer of gum arabic and leave this to dry completely.*

g

Sprayed Crackle

This is a quick way of creating a crackle finish, though the paint does come in quite a limited range of colors. It is a two-part process, requiring a basecoat and a topcoat. The size of the crackle (from small and subtle to big and bold) depends on the thickness of both coats of paint.

1. *Spray the surface with the basecoat, leave it to dry, then spray on a second coat. If you want a heavy crackle, spray another one or two coats.*

2. *When the base coat is completely dry, spray on the topcoat. This can be thick or thin, depending on the crackle you want, but all the top coats must be applied at once without any drying time in between. The cracks will take up to 20 minutes to develop.*

Airbrushing with an Air Gun

This is quite easy to do and you can work on a fairly large scale, though you may need to use several aircans to cover a whole room. When you buy an air gun kit, do read the instructions carefully, as different brands may have different methods of use. You can use emulsion paint, but you must dilute it to the consistency of cream. I used three different colors of emulsion to make these stripes and then sprayed over them very lightly with the background color to soften them.

Pour the paint into the special container and fit the container to the air gun. Press the trigger to start spraying and move the gun along in a slow, smooth action. Avoid spraying the paint on too thickly or it will run.

Airbrushed Pattern

To spray this ceramic dish, I used a special ceramic paint that you can bake in the oven to make it permanent. This was thin enough to use without diluting it. I cut strips of sticky-backed plastic and stuck them onto the dish to mask off a pattern.

Spray the whole dish with the paint. I made the paint thicker at one end to give a graduated effect. With a craft knife, lift one end of each strip of sticky-backed plastic and carefully peel them off. If you have oversprayed onto the underside of the dish, just lift it up from underneath and wipe off the excess paint with a cloth. Follow the manufacturer's instructions for baking the paint.

Airbrushed Shading

An air gun is a good tool for creating a graduated effect. As I was spraying onto a fabric lampshade, I used a thin fabric paint to spray with. I wrapped the lamp base in an old cloth to protect it before I started work.

Spray right around the bottom of the lamp base to make the initial dark area. Then, using a waving-wand action, spray up the shade, allowing the color to get fainter as you move toward the top.

Airbrushing with a Diffuser

These diffusers rely on your breath, so don't try to do a whole room or you will end up with a bad headache. You need to dilute the paint to not much thicker than plain water, or you just won't be able to puff it through the tube. You could make these spots with an airbrush if you are really short of breath.

Dilute the paint with water until it is very thin then dip one end of the diffuser into it. Keeping this end in the paint, blow through the other end to make a spot on the surface. Don't make too heavy a spot or the paint will run; it is better to give it another coat later on.

Monoprinted Fabric

This is the basic, and very simple, technique you use for monoprinting onto any surface. The only difference will be the type of paint you use, which must be suitable for the surface you are going to print onto. Dark colors will show up well on light-colored fabrics, but you may have to mix a light color with white paint for it to show up on dark-colored fabric.

The wonderful thing about monoprinting is that you can experiment with your design on the piece of acetate before committing to printing it on the fabric. If you make a mistake or are not happy with the initial design, just wash it off the acetate and start again.

1. *Pour some pigment fabric paint into a saucer and paint a design onto the acetate with a paintbrush.*

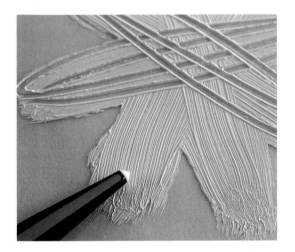

2. *With the handle of the brush, scratch into the paint to enliven the design.*

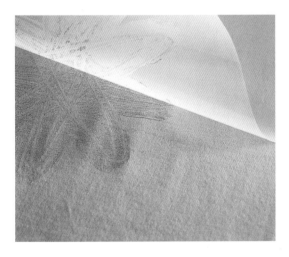

3. *Position the design over the fabric, then lay the acetate in place, paint-side down. Being careful not to move the acetate and so smudge the design, smooth over the back of the painted acetate with your hand. Peel back the acetate to reveal the design transferred onto the fabric.*

Monoprinted Ceramic

Ceramic paints are a wonderful way of adding color to plain china. Use them to print details or all-over patterns to tie in with other colors in a room. These paints can be used on decorative plates, but should not be used on surfaces that will come into contact with food. Most are not dishwasher proof.

Cut a piece of acetate to fit the part of the plate you want to print the design onto. To get this grid pattern, sponge paint onto a piece of radiator grill, then lay the piece of acetate onto the grill to transfer the paint. Peel it off the grill and print onto the plate as described before.

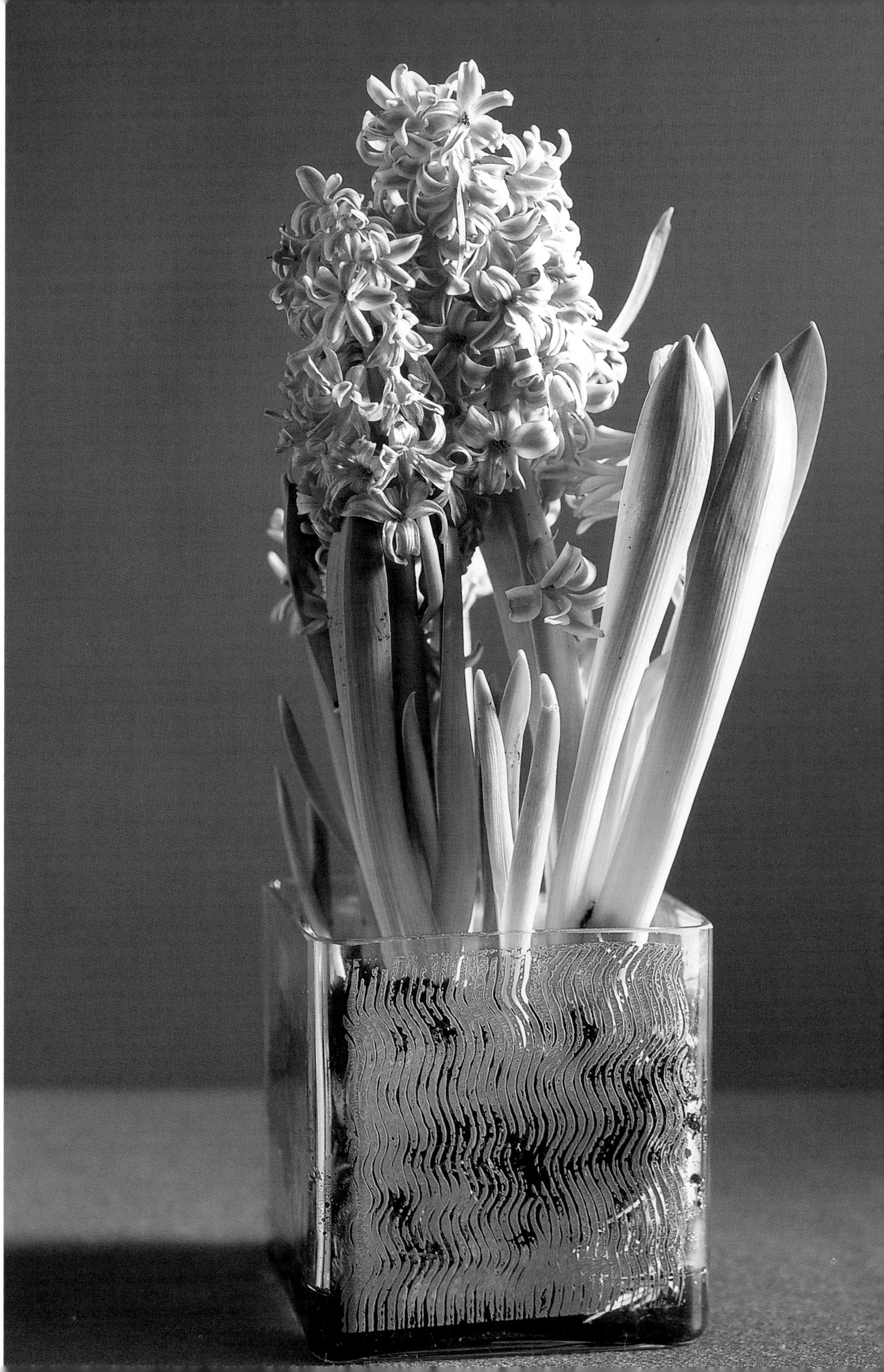

Monoprinted Glass

The reflective quality and translucence of glass make it an ideal surface for monoprinting onto. Choose a color to complement a favorite plant or flowers for a truly harmonious display.

Monoprinted Wall

This is an amazing way to update a decorative scheme quickly and simply. If you choose a semitranslucent pearl paint to print with, you can overlap colors to create a layered effect.

The best base paint is emulsion, though you can print onto other paints; experiment on an unobtrusive corner first.

Cut a piece of acetate to fit the area of the vase you want to print onto. Using a paint brush, cover the acetate in glass or ceramic paint, then scrape a pattern into the paint with a rubber comb. Print onto the glass using the technique described before.

To print squares, simply draw a square of paint onto acetate with a paintbrush—do not apply the paint very thickly however. Swirl and stipple the paint with the brush to create textures, all of which will be picked up in the final print. Print onto the wall using the technique described before.

Painted Stripes

Random, wiggling stripes work well on a feature wall, but it isn't an instant effect. You are painting by hand, so it can be quite slow. However, you can make it as complicated or simple as you want, making the lines dense in some areas and more spacious in others. This finish would work well in a modern room where squarish furniture and clean angles would contrast well with the random lines.

Painted Picture

If you really feel that you cannot draw, you can still use this technique. Either ask an artistic friend to draw out the picture so you can paint over it, or use the projection technique (see page 72) to establish the image.

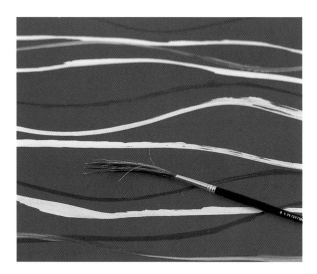

A long-bristled lining brush is excellent for painting a flowing image and you can use a shorter one for the more detailed areas. Try to use flowing, bold strokes rather than cramped ones for a more relaxed, painterly look.

Use a long-bristled lining brush to paint with; they are designed for maintaining a smooth flow of paint over a long line. Simply load the brush with paint, lay it against the wall and draw it down in waving lines. To change the thickness or density of a line, you can rotate the brush in your hand as you pull it down the wall.

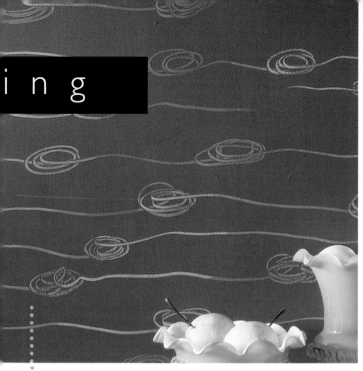

Scraffito

This is a brilliantly easy technique that will take you straight back to the classroom. You can use this technique on a whole wall, but as you have to work quite quickly, it is best to put on the topcoat in small sections at a time. The color of the topcoat will predominate, so it works best if the undercoat colors are quite strong.

1. *Paint on patches of three or four different colors of emulsion as a basecoat. Don't worry about the patches being perfectly painted as they will be almost obliterated by the topcoat. Leave the basecoat to dry overnight. Mix the emulsion for the topcoat half and half with glaze and paint it on across the wall in sections a yard deep.*

2. *To make the scratched lines, I used the eraser on the end of a pencil. Simply drag the eraser through the topcoat in whatever design you choose. You could use an ordinary eraser, cut it into a shape and drag small sections or whole lines with that.*

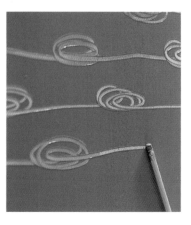

Freehand Monoprint

Monoprinting appears in various guises in this book (see page 86). I used it here to paint stripes onto fabric as this is very difficult to do directly. The stripes tend to end up as rather irregular and blobby. By using acetate to transfer the paint you can create a more professional-looking and more attractive design. I painted a piece of fabric that I then used to cover a notebook.

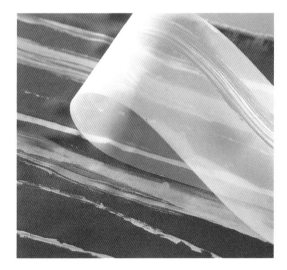

Paint the stripes of fabric paint onto a sheet of acetate with a paintbrush. Lay the acetate paint-side down onto the fabric and smooth over the back of it with your hand. Peel off the acetate and leave the fabric to dry, then iron it on the back to fix the color.

When you are painting a room, it is the way in which you combine the colors and the techniques you have chosen that will make your decorating scheme successful. So, in this final chapter, I have created five very different rooms, each using a variety of paints and effects, to show you how different colors and finishes can be combined for wonderful results.

I have chosen five decorating styles—from eclectic country to urban chic—each of which uses a different color palette and range of techniques. So, whether you favor cool tones in a calm environment or hot hues in a visually wild space, there will be something to inspire you. When you are planning your own scheme, first consider the conditions in the room. How does the light change

throughout the day? What is the room used for? Are there any fixtures, fittings, or furniture that need to be treated? What is the condition of the walls and woodwork? Once you have considered these aspects, you can decide on a color scheme that will enhance the space and choose a range of paints and techniques to suit the surfaces you want to decorate.

If you are unsure about a particular color, buy a small pot of the paint and paint an area close to a window. Leave the paint to dry and look at it at various times of the day to see how the light affects the color. Also, do prepare the room before you start work; cover any furniture and mask off any areas you don't want to paint, otherwise you will spend hours cleaning up after yourself.

I wanted to create an alternative country style in this room—not a dainty look or an overly rustic finish—by using paint to add rich color and texture. I wanted the room to be relaxing and feel comfortable, but also to be stimulating with lots to look at. Each surface required a different treatment using a variety of paints and techniques.

The paint colors were chosen by looking at what was staying in the room, which was a large sideboard that rather dominated the space and some fabulous leather cubes and cushions.

Eggplant-colored textured plaster was applied to a panel; iron paint was patinated to give a rusty effect; the oak wall added a warm honey glow to the space; while the metal-painted panels gave small areas of accent colors that picked up other colors in the room and took them up to eye-level to give a coherent feel.

To enhance the look, I added warm colors and sensual fabrics. The suede and leather furniture not only feels wonderful but gives structure without stiffness, as it has straight but soft edges. One of my favorite pastimes is styling my home, rearranging the furniture and ornaments; for this room I grouped collections of vases and chose strong, simple pieces for an unfussy, bold effect.

Since I was a child, I have had to touch, in order to understand a surface, much to the horror of my parents when we entered glass and china stores. I'm very much drawn into this room and enjoy spending time in it, because each surface has its own identity and differing qualities that make you want to reach out to touch and explore it.

Rusty Plinth

I wanted to make this innocuous **MDF** plinth look and feel like metal so
I chose a metallic paint with a difference. This rusty iron paint contains
metal, so it is heavier than normal paints and is quite smelly, but to
achieve this rusted effect was simplicity itself (see page 16).

   You must seal the MDF before you patinate it and I used black emulsion
paint to do this, as the rusty paint is black when you first paint it on. As I
wanted a really rusty look, I used quite a lot of the aging solution. On the
vertical surfaces this works spectacularly well, as it runs down the surface
creating really natural-looking rusty drips and smears. The rust patina
contains an mild acid solution, which sounds hazardous, but as long as you
are careful and follow the instructions this is one of the easiest techniques
to achieve and needs no special skills at all.

Patinated Panels

These three pieces of patinated MDF were
attached to the wall with heavy-duty touch-
and-peel pads and are still holding strong.
They look horribly complicated at first
glance but in fact they are simple to
achieve and are done in the same way as
the plinth on the left. To seal the MDF and
provide a good base color, I painted each
panel with an appropriately colored
emulsion first. I then painted the panels
with reactive paints (see page 16), which
contain metal and which once dry, can be
sprayed with aging solution to achieve
black, green and blue patinas that simulate
a variety of aged metal effects.

Polished Plaster Panel

The plaster panel on the right was great fun and quite addictive to do (see page 36). Once again, it is all about preparation, masking off the area and laying down cloths to protect the floor. I added acrylic colors to Venetian plaster and mixed it well with a spoon. Venetian plaster is thick and it is hardwork to mix the color thoroughly, but it is important to spend time on this. Venetian plaster comes in light, medium, and dark bases; the darker the base, the deeper the final color. I used a dark base as it contains less white pigment and so less acrylic color is required. This technique would be ideal if your walls are lumpy or damaged from scraping off old wallpaper, or if you are like me and want to satisfy your thirst for texture.

Dragged Wall

Dragging is normally used vertically (see page 56), but I decided to turn that idea on its head and see what the effect would look like done horizontally. I think it's my rebellious streak coming out!

The most important part of this technique is preparing and masking the walls, ceiling and skirting. With this done, the painting takes no time at all. I used a taupe-pink emulsion as a background color, which I left to dry overnight, and mixed emulsion with scumble glaze for the topcoat. Starting at the top of the wall and using a brush, I painted a yard-wide strip of emulsion-glaze right across the wall. With a wide dragging brush, I dragged across the wall horizontally. You will probably have to work in sections at the top of the wall, moving the ladder as you go. Where this is the case, blend in the beginning and end of each dragged section by gently lifting the brush as you reach the end of the stroke and then feathering the end of the next stroke in the same way so that the two strokes blend together. This may take a little practice to do well, but if you make a mistake, just paint the emulsion-glaze smooth again and have another go.

Woodgrain Wall

This faux-oak effect (see page 68) is one of my favorite techniques because it is easy to pick up and the end result is impressive. It also works very well on a large area and so it's perfect for a whole wall. You normally find oak in halls where it is paneled and darkened with wood stain, but I didn't want everything in this room to be strong in color and texture so I used the natural color of un-treated oak.

Woodgrain Panel

If woodgraining a whole wall seems a little daunting, then you could try treating a small panel. You can lay the panel flat, which makes it easier to work on, and when it is dry, hang it on the wall as an artwork. I used the knotted wood technique here (see page 68), but you could use any of the woodgraining techniques.

Painted Sideboard

I had a nice piece of furniture that I found very useful as it provided lots of storage but unfortunately it had some wear and tear on the melamine top. I had tried arranging the ornaments to cover all of the chips, but it ended up looking rather odd.

Painting melamine can pose a problem because most paints are unsuitable for it; emulsion paint would stick at first, but in a short time it would flake off. The answer is to use specialized melamine paint (see page 30), or if you want to use a particular color unavailable in the melamine range, use a melamine primer, over which you can use emulsion.

I primed the surface with two coats of melamine primer using a short-pile roller and leaving it to dry in between coats. I mixed metallic paint and emulsion together, half-and-half, to get the color I wanted and roll it over the primer. Then I applied two coats of a good-quality matte acrylic varnish so that the paint would remain beautiful.

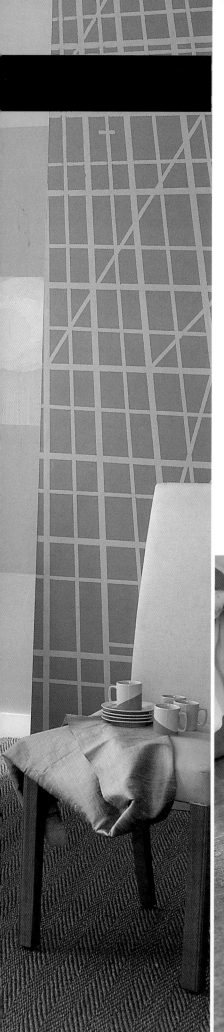

When I started planning this room, the walls were completely blank, which can be daunting at first, so I spent some time visualizing how I wanted the room to look. This room is a structurally driven space, making an exciting, contemporary dining room. However, I wanted it to feel grown-up and elegant, so the colors had to have a classical, timeless quality, but not be at all old-fashioned. So, I chose a palette of soft taupes, creams, and khakis.

I chose the colors for the squares and circles wall with special care. I did not want them to jar or clash in any way. It can be hard deciding upon various colors to be used side by side but the reason that this wall is so successful is because I decided upon three colors and then mixed them together to make a total of six. This method results in a group of colors that sit very happily together.

The geometry of the squares on the back wall was complemented by using three sizes of circles, which were made by tracing around three different-sized plates and then painted with metallic paints to add extra luster. The perpendicular lines of the furniture complement and become an extension of the wall finish.

Mosaic Table Runner

I used a piece of glass ¼ in.(6 mm) thick with polished edges. Clean the glass thoroughly and lay it face down on a towel to cushion it. I used a sponge stamp (see page 80); the only difference being that I used acrylic size instead of paint to stamp with. I poured size onto a plate and gently pressed the sponge stamp into it, then dabbed off the excess. I pressed the stamp onto the back of the glass and then lifted off and repositioned it. If you make a mistake, simply wipe the size away with a damp cloth, leave it to dry and start again. The more size you put on, the more gold flakes will stick to it. I used a small amount of size to give quite a subtle effect.

When the size is tacky to the touch it is time to sprinkle the flakes of Dutch gold over the surface (see page 44)—not much is needed as it goes a long way. I pressed them on and dusted away any excess. As the mosaic is on the back of the glass, it will not be damaged by hot dishes.

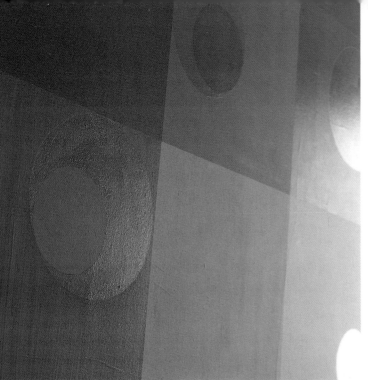

Squares-and-Circles Wall

Masking (see page 76) a wall into blocks of color takes
patience and time, but the results are stunning. To achieve
this effect, you have to measure the height of the wall and
divide it by the number of blocks you want. Mark the edges of
the blocks at the ceiling and skirting with a pencil. The only
thing to bear in mind is that because the tape lies over some
squares you can paint only alternate blocks in one session.
You must wait for the paint to dry before repositioning the
tape and starting to paint again. On some of the painted
circles, I then painted another circle in one of the wall colors,
adding to the layered effect.

    If these colors are not to your taste, why not change them
to inky blues and purples, or hot pinks and fiery reds, or
whites and silky creams?

Découpaged and Gilded Wall

I covered the wall on the right in
crushed paper (see page 70) but
instead of using fabric dye to color
the paper, I used black permanent
ink. I glued the paper in place and
left it to dry overnight. For the
gilded squares I used a metallic foil
(see page 24) that comes in the
most wonderful colors; I have used a
soft gold/copper. I used the
patchwork gilding technique (see
page 64) to apply two coats of aqua
size and then applied the film.

Masked Wall

This painted wall was inspired by a picture of scaffolding, which may sound bizarre, but inspiration can come in many guises. I wanted to create the illusion of perspective and worked carefully to recreate the scaffolding image. I painted the background in a very pale gray and left it to dry overnight. I masked off the lines (see page 76) with two widths of low-tack tape, firmly stuck down to stop paint seeping underneath it. I then mixed darker gray emulsion with glaze (see page 20) and rolled on two coats, allowing it to dry between coats. When the paint was dry, I peeled-off the tape.

An alternative idea, which has been tried and tested by me, is to mask off a rectangle on a wall. Create your masked artwork, then paint the whole wall in a slightly darker color, allow it to dry and remove the tape. The end result is a framed artwork for the price of a roll of tape and left-over paint. A great way to utilize a good-colored but grubby wall that needs repainting.

Dragged Wall

I love the quality of flat matte paint when it is accompanied by texture. I also wanted to add a softness, which dragging does so well (see page 56). I got this result by dragging a square-headed stippling brush, studded with soft rubber bristles, through the paint. The brush was not designed for that purpose, but if it works, then use it!

Urban Living Room

I designed this cool, fresh living space to complement the dining room shown on the previous pages. I wanted the room to feel relaxed and informal, so I chose a color palette of soft blues and pinks, with a touch of green.

To create the design on the walls, I used exactly the same masking technique as I did for the squares-and-circles wall (see page 102). However, instead of masking the whole area into regular squares, I masked off a number of different-sized squares and rectangles across the walls. This was easy to do as, on the whole, the squares don't touch, so the whole area could be done in one go. Where the squares overlap, I masked and painted the one in front first and when it was dry, masked and painted the one behind.

I stenciled (see page 44) the grids of gilded circles onto the walls once the paint was dry. I stippled size through a cardboard stencil and when that was tacky, I gilded (see page 64) onto it with silver leaf.

This room attempts to bring the outdoors inside by bringing colors and shapes from the garden into the conservatory. This has been done both literally, by using real leaves to stamp a design, and visually, by utilizing a palette of gentle blues and soft greens. The whole theme makes this room a pleasurable environment where one can feel calm and cool—a place to reflect and unwind.

At first glance, this conservatory looks white, but, in fact, every surface has been painted a color; a light tone of color, but nonetheless a color.

I love to combine the quality of flat matte paint with texture and this room really does show how well that can work. Nothing is overbearing; there are no discordant notes or jarring textures. Imagine retiring to this room on a hot day to sit and let your mind wander or immerse yourself in a good book, accompanied by a long, cool drink. Perfect.

Painted Vase

The variety of specialized paints (see page 30) available is becoming increasingly interesting and there are different types to cover a multitude of surfaces, including ceramic, which is what I have painted here. This particular vase is porous, so I chose a soft, fine paintbrush to get the detail I wanted in my painting. I drew the design out in pencil, though you could, of course, trace it if you are not confident about drawing. Then I painted the picture, starting from the top of the vase to prevent smudging what I had already done.

Artwork Panel

The panel below of four ferns was achieved by reproducing the same image in a variety of ways. The board was divided into four upright rectangles. Starting from the top, the first image is a colored transfer (see page 72). The next image was done in the exactly the same way, but I reversed the photocopy to give the negative colors. The third image was made using the piece I had cut out of the stencil I used for the relief wall stencil on the right. I positioned the fern shape on the panel with a few pieces of rolled-up tape underneath it to stop it from moving. I masked off an area around the fern shape with tape and covered it with gesso, completely covering the fern shape. I immediately removed the fern shape and the tape and left the gesso to dry. The final fern was a color photocopy of the fern which I cut out and glued in place with diluted white glue (see page 70).

Gesso-stenciled Leaves

To make the stencil used on the wall above, I traced around a real fern and cut out an acetate stencil with a hot cutter (see page 44) to achieve this rather larger than life fern. I melted the gesso to the consistency of paste and taped the stencil in place. I simply troweled the gesso on, filling the stencil (see page 44). Remove and reposition the stencil across the wall until you have completed your design.

Woodgrained Table and Stool

Believe it or not, doing this knotted wood technique (see page 68) is quite infectious and when we photographed it for this book everyone had to have a go, because it is quite magical. From this strange-looking piece of rubber comes a wonderful effect. I wanted to break with convention and decided to woodgrain the furniture in soft and gentle colors so it sat happily in its surroundings. I chose a white undercoat and a pale sage-green topcoat.

Stamped Leaves

I find this technique (see page 48) very satisfying because it makes a very small amount of paint go a very long way. The same effect could be achieved with ferns or feathers. I rotated the leaves so that the tip of each leaf did not go in the same direction, giving a more natural appearance, and used emulsion paint to stamp with.

Combed Wall

I painted the whole wall with pale gray emulsion and when it was dry, I mixed a glaze with acrylic color (see page 20) to make a creamy-brown topcoat. I painted on the topcoat in yard-wide sections and combed it with cardboard (see page 50). Starting from the top, I simply dragged the cardboard down the wall in one swoop. This can take a few goes to get right, but that's all right because the glaze gives you an extended working time, so paint over it again and repeat the process until you are happy. I have added an extra element to this wall by combing over small areas with an oak grain comb (see page 50) but this is optional. If you want to do this, start at the top of the wall and work down, working with the grain of the cardboard combing.

Grid-pattern Floor

I had thin ⅛ in. (3 mm) thick **MDF** cut to size then glued and pinned in place. I filled all the gaps and painted it with two coats of soft gray-green emulsion. I masked off (see page 76) a border around the edge of the panels and painted it a dark metallic olive-green using a short-pile roller. The gridded pattern was then masked off and painted with silver paint, which I stippled onto the masked stripes to give them texture. The whole floor was varnished several times to protect the surface.

Feather Radiator Cover

I found Lasertran, a new and exciting product (see page 73) that can be used on a multitude of surfaces, even toast apparently. I stretched the fabric taut and stapled it in place within the wooden radiator cover. I then primed the fabric with acrylic primer and when that was dry, painted the areas I wanted to decorate with two coats of diluted white glue before transferring the image.

An alternative idea would be to put a portrait of each member of your family onto the back of an upholstered dining chair; the only limitation with this technique is your imagination.

Our design history is steeped in floral patterns; for centuries we have decorated our homes with their shapes and colors, in a multitude of designs. Their familiarity, however, means that the images can become stale and dated. How can we reinvent something that has been designed and redesigned for centuries? Simply by changing the scale and the colors.

I chose spring green as a background color for the walls, with over-sized flower designs to give an extravagant feel. The contrasts of the soft and bright colors on the headboard and linen reflect the contrasts in nature.

This room had to be fresh and relaxing, a place where I could spend time. I combined the flower motif with a variety of techniques and paints to create a truly inviting space.

Crackled Plinth

I used a basecoat of fresh lemon yellow and a crackled topcoat of leaf green to give a new lease of life to a traditional effect (see page 82). I basecoated the plinth, then crackled one side at a time, allowing each to dry before doing the next one. I did the top last of all.

Frottaged Headboard

I divided the headboard into eighteen pieces, with a small gap between each one, and cut paper to size. I basecoated it with gray emulsion and when it was dry, rolled on a topcoat of lilac pearlized paint Starting with the top left-hand corner, I frottaged (see page 54) across and down to the bottom right-hand corner.

mian

Projected Eucalyptus

I wanted a three-dimensional feel, with the eucalyptus falling across the wall and onto the headboard. I placed sprigs of eucalyptus directly onto the projector (see page 72) and moved them until I was happy with the positioning. Then I painted the leaves with an artist's paintbrush, using purple pearlized paint, a section at a time so that I could frottage (see page 54) the image as I worked.

Projected Mimosa

To create the image I arranged, then photographed, sprigs of mimosa. I took the photograph to a copying shop and had it copied onto acetate, which I then laid on the projector. Once I was happy with the positioning of the projected image, I painted over it in soft, oyster-colored pearl paint.

Stippled Border

I basecoated the walls with soft green emulsion paint. I masked off a border (see page 76), then stippled on the shading (see page 74), using a lime-green emulsion paint. These defining lines are a strong background for the projected images.

The Bed

The bed takes up a large area of the bedroom, so be sure to include it in the general design scheme. A more personal, and cheaper, alternative to buying new bed linen is to paint what you already have. Also, you can make exactly the right linen for your new room, rather than having to compromise. I wanted to create a floating design on the linen, not a set pattern that completely covered it. So I concentrated on painting one corner of the duvet with a lot of flowers, while the remainder has a light sprinkling of them. The pillowcases were each embellished with a single flower.

Painted Linen

Painting and batik give an interesting combination of color controlled by wax, contrasted with areas where color has been allowed to bleed for a softer look. I painted flowers (see page 76 and 124 for template) with hot wax, then colored them with jewel-bright fabric dyes.

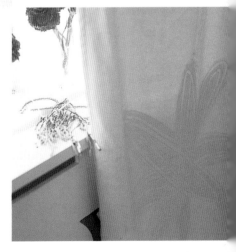

Monoprinted Curtain

I used ready-made cotton velvet curtains to monoprint onto. This fabric has a short pile, making it perfect for printing as you can still appreciate the texture. I laid the canary-yellow fabric on a flat worksurface and monoprinted (see page 86) it with green fabric paint (see page 26).

THE LOOK:
clash it

This is a wild, visually stimulating workplace, but as with everything you do, the concept can be explored and altered to meet your needs. I used emulsions and glitter paints to make the stripes, which were masked off at different widths. However, you could make this room very masculine by simply painting the color stripes in a variety of blues.

The striped wall dominates this room and that was my intention. I wanted it to have its own identity, to feel different from any other room in my home. I work and live in the same location, so making a firm separation of work space from living space is important to me. When I come in to this room I think, "Now It's work time."

Storage and extra surfaces to spread out on had to be included in the design, and I also needed to have my inspiration around me in order to create new ideas.

I loved the look of the vertically striped wall next to the horizontally striped drawers. I would be in a mess without my filing drawers; they give an air of organization by containing all of my stationery bits that otherwise seem to float around and create mess. This room has the right energetic atmosphere that I need to be able to settle down and do some hard work.

Sprayed Blind
Graffiti artists, eat your heart out. I covered the floor with an old sheet then laid a ready-made blind over it and sprayed colored enamel spray paint (see page 18) in a steady stream across the whole blind. I left it to dry then stapled the feather trim to the wooden batten at the bottom of the blind. It takes ten minutes to make this blind; I love jobs that are so quick.

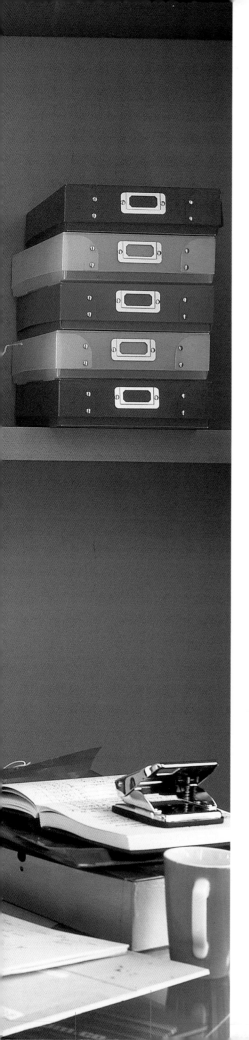

Home office

To offset the glitter wall, the other walls in the room were painted with wider stripes separated with a regular band of dark purple. The stripes were carefully calculated to line up with the edges of the "floating" shelves. When you are painting stripes, it is important to plan them around any fittings in the room. If they sit uncomfortably around a fireplace or cupboard, the strong vertical lines look disruptive and untidy and the whole effect is spoiled. It is best to make the stripes fit in with the fittings, measuring them out and marking them in pencil, adjusting the widths until everything looks right.

Gilded Glass Tabletop

This technique can easily be achieved on glass or a wall (see pages 42 and 64), both of which I have tried and it looks completely fantastic when lit either by daylight or artificial light. It reminds me of the excitement of fireworks exploding on a clear winter night. I painted this table with satinwood paint and then splatter-gilded the back of the glass top, so that the gilding was protected by the glass itself.

Striped Wall

I used masking (see page 76) to achieve the perfect lines of color shown above; they look rather like a multicolored barcode. To prevent it from being visually confusing, I painted one wall with a complicated sequence of stripes, where I butted up bands of different colors, some of which had glitter paint applied over the top of the emulsion once it was dry. I used a low-tack tape to mask the stripes, which lightly adheres to the wall without removing most of the paint once you have finished.

The only skill you need to achieve this wall is to be able to see a straight line. I struggle a bit with this, so I stand back a little distance from the wall and it becomes a great deal easier to see how straight my lines are.

Filing Drawers

The cabinet drawers are made from wood, so I painted them with the same emulsion (see page 10) I used for the walls. The color relationship makes the drawers an integral part of the scheme. To keep the drawers looking nice, I applied a coat of water-based varnish.

Sampleboards

A washing line of tension wire with clips threaded onto it is an excellent way to display samples, mood boards and those all important documents that mustn't be forgotten about. Hanging among the swatches are a stamped fern (see page 48) and a board of rough plaster (see page 36).

Pinboards

These simple cork pinboards are invaluable to me; I have loads of them and pin fabric swatches, tear sheets, and bits of trimmings to them. I just laid the cork tile on a cutting mat and positioned a dinner plate on top, then cut around it with a scalpel. I painted the cork (see page 30) and left it to dry. I attached the pinboards to the wall with heavy-duty adhesive touch-and-peel tape. What could be simpler?

t e m p l a t e s

Stenciled fabric
page 46

Painted linen
page 117

Potato prints
page 48

Stenciled metal flakes
page 46

Acetate stencil
page 45

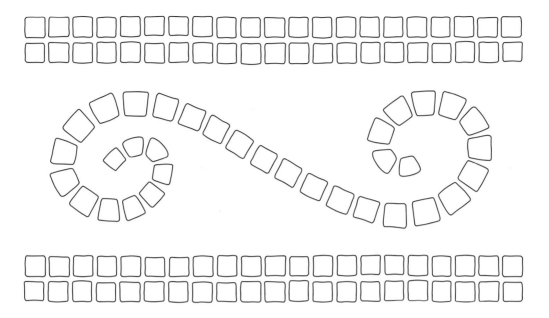

Stenciled mosaic
page 80

Card stencil
page 44

Suppliers

The Art Store
4004 Hillsboro Pike
Nashville, TN 37215
Tel: (800) 999-4601
www.artstoreplus.com
Good source of glass paints.

Benjamin Moore & Co.
51 Chestnut Ridge Road
Montvale, NJ 07645
Tel: (800) 344-0400
www.benjaminmoore.com
Interior and exterior paints.

**Chesapeake Ceramic
Supply Inc.**
4706 Benson Avenue
Baltimore, MD 21227
Tel: (800) 962-9655
Fax: (410) 247-1708
www.ceramicsupply.com
Ceramic colors.

Chroma Inc.*
205 Bucky Drive
Lititz, PA 17543
Tel: (717) 626-8866
Fax: (717) 626-9292
www.chromaonline.com
*A source for classes,paints,
and other useful links.*

Dove Brushes & Tools
1849 Oakmont Avenue
Tarpon Springs, FL 34689
Tel: (800) 334-DOVE
Fax: (727) 934-1142
www.dovebrushes.com

Dutch Boy
www.dutchboy.com
*Clear color charts for
interior and exterior paints.*

Flex Bon Paints
2131 Andrea Lane
Fort Myers, FL 33912
Tel: (941) 489-2332
Fax: (941) 433-0203
www.flexbon.com
*A variety of wood stains,
waxes, and varnishes.*

**G & S Dye and
Accessories Ltd.**
250 Dundas St. W.
Unit #8, Toronto,
ON, M5T 2Z5, Canada
www.gsdye.com
*Everything from fabric dyes
and paints to marbling
colors and paintbrushes.*

Golden Artist Colors
188 Bell Road
New Berlin, NY 13411
Tel: (888) 397-2468
Fax: (607) 874-6767
www.goldenpaints.com
*An artist destination for
acrylic paints.*

Howard Product, Inc.
560 Linne Rd.
Paso Robles, CA 93446
Tel: (800) 266-9545

Fax: (805) 227-1007
www.howardproducts.com
*Wood care products,
polishes, finishes, and waxes.*

Loew-Cornell, Inc.
536 Chestnut Ave.
Teaneck, NJ 07666
Fax: (201) 836-8110
*A variety of brushes and
transfers.*

**Mann's Woodcare &
Restoration Products**
P.O. Box 343
East Winthrop, ME 04343
Tel: (207) 395-2739
Fax: (810) 283-2465
briwax.safeshopper.com
*The place to find liming
and natural wax.*

Modern Options
www.modernoptions.com
Tel: (800) 447-8192
Fax: (800) 221-6845
*The source for metallic
paints.*

**The Painter's
Workshop**
www.paintersworkshop.com
*This site is full of useful
things, such as classes,
books, materials and other
great link all dealing with
painting.*

Prismflex
1325 Eddy-Scant City Rd.
Arab, AL 35016
Tel: (800) 240-8787
www.prismflex.com
*Suppliers of screen
painting inks.*

Progress Paint
Manufacturing Co. Inc.
201 E. Market Street
Louisville, KY 40202
Tel: (800) 626-6470
www.progresspaint.com
*Easy to use chart to find
the right type of paint for
any surface.*

**Wagner Spray Tech
Corporation**
1770 Fernbrook Lane
Minneapolis, MN 55227
Fax: (612) 519-3563
www.wagnerspraytech.com
Roller assorts available.

Texas Art Supply
2001 Montrose Boulevard
Houston, TX 77006
Tel: (713) 526-5221
www.texasart.com
*Great place for supplies.
Everything from brushes
and dyes to batik fabrics
and other painting supplies.
Not a place to be missed.*

Author's Acknowledgments

Thanks to the team of amazing people who have made all this possible. It has been a pleasure to work with each and every one of you. Thank you firstly for being my friends and secondly for being incredible professionals.

Kate Haxell, for remaining sane while editing my version of the English language! You have been the backbone, keeping us all individually on track.

Cindy Richards and Mark Collins for having the vision and belief to make all this a reality. Robin Gurdon for being a gentleman in its truest sense. Lucinda Symons, for your love of it and extraordinary vision. Holly & Emma, for your diligent work throughout what sometimes bordered on chaos.

Lucynia Moodie for your passion and unrelenting inspiration. Robert Atherton for sharing my maddest painting moments. You're a great pal. Janet James, for your enormous amount of patience and love you've invested in making this book one that I'm immensely proud of. You're a star.

To all those special people in my life who have really made a difference.

Mum, who has, from time to time, been found up a ladder. My sister, Juliet. Tim Killingbeck for your friendship that I hold so dear. Doo Gurney for your unbreakable resolve—I watch and learn. Daniel Sims. My Auntie Rene. Astra and Alan Minchin. Karen, Paul, and Jessica Wheeler. Vicky and Peter Farren. Peter Mooney. Debbie Pound. Joanne Barnet and Andrew Kalinowski. Anthony, Sarah Alfred, Missy and my godson, Leonardo De Signey. Simon Laity. Sam Henwood. Sarah, Peter, Katie and Jemma Robbins. Avril Graff. Julian Ionergan. Paula Mattin. Sally and Phil Wren. Jim Alabaster and my business guru Paul O'Byrne.

I would also like to thank the following for their help and for generously giving me materials for use in this book.

Amo and Jane Vich at E Ploton for their constant support. Emily Sears at Cuprinol for garden and wall shades. Juliet Strachan at Mode Information Ltd. Mala Sarwal at Artex-Blue Hawk Ltd. Mark Gower at Gordon Audio Visual. Mick Hawkey at H2KFX. Plasti-Kote for crackling base and top color and Fleckstone. Polyvine for colorizers and glazes. Richard Burbidge for skirting boards. Shelton Kartun at Oakthrift Plc for the paintstick. Steve McGilveray at Craig & Rose. The English Stamp Company for stamps and paint. Veronica Johanffon at Home Key. Don Lockwood at International color Authority. Barbara Jones at The Mix. Sarah Harrison in connection with International Ltd.

Publisher's Acknowledgments

Cima Books would like to thank the following companies for loaning items for photography.

Kersaint Cobb, +44 (0)1675 430430 natural flooring.
Keramica, +44 (0)1782 207206, ceramics.
Amtico +44 (0)800 667766, flooring.
Bill Amberg, +44 (0)20 7727 3560, leather goods.
Goldfinger, +44 (0)20 8694 2603, furniture.
Brintons, +44 (0)1562 820000, carpet.

Index